THE UNFAIR ADVANTAGE

— · —

HOW SMALL BUSINESS OWNERS CAN USE ARTIFICIAL INTELLIGENCE (A.I.) TO BOOST SALES, OUTSMART THE COMPETITION AND GROW THEIR DREAM BUSINESSES WITHOUT BREAKING THE BANK

PHILIP BLACKETT

PREFACE

"In the long history of humankind (and animal kind, too) those who learned to collaborate and improvise most effectively have prevailed." - Charles Darwin

E mbarking on this literary journey, it is my intention to nurture a space where collaboration and innovation not only thrive but set the stage for small businesses to achieve groundbreaking success. Cutting through the trepidation that shrouds Artificial Intelligence, *The Unfair Advantage: How Small Business Owners can Use Artificial Intelligence (A.I.) to Boost Sales, Outsmart the Competition and Profitably Grow their Dream Businesses without Breaking the Bank* has been meticulously crafted with a heart to serve the entrepreneurial spirit that yearns for guidance and a mind that seeks tangible results for small businesses globally during such changing times revolving around new trends and advanced technologies.

In an era teeming with technological advancement, it is crucial that we grasp the tools available to grow our endeavors. Spawned from a deep-rooted passion to shepherd small business owners to not just exist but excel in an increasingly competitive landscape, this book is a beacon that shines on the practical applications of AI, tailored to meet the unique needs of entrepreneurs, startups and small businesses often seen as the backbone of the American economy.

It is, therefore, my mission to demystify AI, turning perceived complexity into a wellspring of opportunity, demonstrating that using AI does not necessitate a behemoth budget, and proving that AI has its place in the arsenal of tools for businesses of all sizes. Now, with that being said, this book will take more of a top-level approach addressing mindset and strategy, given that the actual A.I. tools that we can use may change as quickly as by the time you turn the page. In any case, I will provide a way for us to stay connected where I can share with you the top tools that I am using and am recommending for the small business clients that I work with each day.

Now, reflect upon the tenacious shop owner whose eyes sparkle with the joy of building relationships with every customer yet loses sleep over the relentless pressure of competition. Envision the solopreneur whose perseverance has been the cornerstone of his or her success, yet he or she can't shake the feeling of potentially being left behind in a digitalized world. These narratives echo the voices of many who stand at the threshold of integrating groundbreaking technology into their livelihoods.

These are the voices that beckoned me to humbly write, to offer solace, direction, and confidence in harnessing AI's transformative potential not only to change the game in how it is played but to enhance the odds of winning for those who take AI seriously and learn how to adopt it and utilize it in growing their businesses sooner than later.

Inspiration has poured in from ancient scriptures, past experiences in business, seasoned wisdom from the realm of politics, and robust theories from the annals of economics. Their collective insights weave and intertwine through these pages, offering a diverse tapestry of knowledge that I hope is helpful as you explore what A.I. is and how to use it to grow your business.

I humbly extend my gratitude to the mentors, colleagues, and, most importantly, the small business community whose timeless stories and challenges have been the motivation and the lifeblood of this work. A special shoutout to all my readers of my *Use AI to Grow Your Business* newsletter on LinkedIn who have regularly read my weekly posts on how artificial intelligence can be used to grow their businesses and careers. You all played a role in me stepping up to write a book like this to better serve you and for those who are not yet readers of the newsletter globally. A special note of thanks to my loving family, whose unwavering support has been my anchor and to my Lord and Savior Jesus Christ, whose teachings serve as the compass guiding my path here with you today.

To those who have chosen this book, I recognize the sacred ground upon which I stand – your time, trust, and eagerness to grow. You are the diligent stewards of your endeavors, the driven architects of your dreams, and it is a privilege to share this part of your journey. I am humbled and grateful for your attention, and I will do my best to give you the best possible return on investment for your time reading this.

This work beckons those at the helm of small businesses – from the spirited innovators to the seasoned veterans of commerce – no prior knowledge of AI is required, only the desire to flourish and an openness to embrace new paradigms and strategies.

In appreciation for your purchase, I invite you to continue reading. Let us embark on this transformative journey together, unfurling the sails of curiosity and anchoring in the safe harbor of knowledge. As you turn each page, may you find the solutions you seek, and together, we will forge a future where your business not only survives but thrives.

With warmth and anticipation for what awaits,

Philip Blackett

P.S. You don't have to wait until the end of the book for me to invite you to consider working with me personally.

If you are open to the concept of working together with me first-hand to use artificial intelligence to grow your small business, visit my website at www.DreamBusinessMakeover.com and schedule a time so we can discuss the opportunity of accelerating your growth together.

CONTENTS

1

AI DEMYSTIFIED: ENABLING THE SMALL BUSINESS ADVANTAGE

In the sunlit warmth of an early afternoon, Sarah's plants cast a number of shadows across the ledger that sprawled open on her antique oak desk. Worry lines creased her brow as she navigated the budgets of her small boutique — a labor of love, born from a mix of daring and necessity. Amidst the rustle of ledger pages, the gentle chime of the storefront bell announced a customer's arrival, yet her thoughts remained ensnared elsewhere — in the realm of potential and progress.

Artificial Intelligence (AI) — this concept that once felt distant especially for smaller businesses like hers, like starlight from millennia ago, suddenly felt close enough to touch and warm enough to cast its own shadows within her quiet dreams. She had heard it whispered in the corridors of her local gathering of business owners, had seen it wielded by corporate giants on popular business news TV programs, with its very name evoking a symphony of productivity and precision she ached to conduct.

The customer's voice — a soft, inquiring murmur about a vintage shawl — pulled her back to the immediacy of service, to the tactile world of textiles and tender transactions. Sarah's smile was a practiced art, a testament to her resolve and resilience amidst the clamor for change. She navigated these interactions with an air akin to dances she had seen on old film reels — elegant, purposeful, but always following a familiar rhythm.

The balance of her days was a mosaic of such moments — threads of the mundane woven with strands of aspiration. She envisioned a future where the gentle hum of AI music would pervade her operations, where the data of countless transactions became a chorus in a symphony of patterns, beheld and interpreted with divine insight, casting revelations upon her strategies like sunlight through stained glass.

By twilight, the boutique whispered goodbyes to the last patron, and Sarah stood alone once more. She reached for the ledger, its numbers a testament to her toil, and wondered — What if these rows and columns could be strewn across the canvass of algorithms, each entry a brushstroke of possibility in the grand design of business growth and customer fulfillment? She imagined a world where the hands of time ticked to the rhythm of efficiency, where AI was not an oracle seated atop an inaccessible peak but a companion on the journey along the road less traveled by small business owners like herself.

And as she turned the sign on her front door to "Closed," her mind, that crucible of both fears and frontiers, posed a question to the approaching night: Could her small boutique, with its local roots and artisan heart, find sanctuary and strength in the embrace of artificial intelligence, blooming into a transformation as profound as scripture promises to an open and faithful heart?

The Dawn of AI in the Entrepreneur's World

Artificial Intelligence (AI) has transcended the realm of Silicon Valley boardrooms and now knocks on the door of the local entrepreneur, offering not just a toolkit but a partnership to scale the steep slopes of modern commerce. The age-old narrative that positions AI as the behemoth's playground is not just outdated—it's a misconception that strips smaller ventures of their rightful claim to innovation. Here in the inaugural chapter, we chart a course through the mist of uncertainty that shrouds AI, laying a path that is not only accessible but remarkably advantageous for the small business owner.

The democratization of AI is upon us. No longer squirreled away behind the monolithic walls of tech giants, AI beckons small business owners to embrace its potential. The gifts that AI gives are now dispersed for the upliftment of all businesses, big and small. That is, for all businesses who take advantage of learning and adopting it now. It stands as a testament to our collective progress that technology which once required the resources of a Goliath can now be wielded by David.

Small businesses are often the heart and soul of our communities. Artificial Intelligence now serves as the whetstone, allowing small enterprises to sharpen their competitive edge. Through this transformative technology, these businesses can automate mundane tasks, analyze data with a precision unattainable by human hands, and personalize customer interactions. This is not just an operational upgrade but a strategic revolution.

Empowering the underdog, this chapter provides a robust foundation for understanding AI's relevance and application in the small business ecosystem. It is a call to arms for the resilient, the dreamers, the entrepreneurs who gaze upon the technological horizon and yearn to harness the winds of change. Let this not be a tale of 'what if', but a chronicle of 'what is', as AI becomes an integral ally in the pursuit of exponential business growth, enlarged missions to fulfill, and financial prosperity that can lead to generational wealth for families globally.

What then are the tangible benefits for the small business owner? AI deftly cuts through the Gordian knot of inflated operational costs, bloated labor budgets, and inconsistent service deliveries. It simultaneously enhances customer experiences, creating a mosaic of interactions that are both more personal and more efficient. The loaves and fishes multiply as AI allows for the optimization of scarce resources, feeding the ambitions of growing enterprises with opportunities previously unimagined. With AI, small businesses can afford to do what corporate giants do all the time: do more with less.

This chapter will illuminate three pivotal insights: the myth-shattering reality of AI's accessibility for small businesses, the transformative impact it can have across various business functions, and the profound capability of AI to profitably reduce costs and elevate customer experiences. Picture a craftsman's workshop, where every tool has its place and purpose; so too does AI find its home in the small business, a versatile instrument in the craftsman's hand. That would be your hand.

We embark now on a journey that reveals how mastery of AI can bring about a new epoch for the small business owner. You will be privy to practical strategies and real-world examples that not only elucidate the power of AI but instruct you on its application. We forge ahead, confident that the insights gained will open a gateway to growth and unparalleled innovation—for it is in the marriage of divine purpose and human insight that businesses are not just built, but they thrive.

Harnessing the Momentum of Artificial Intelligence

Venture forward with us as we dissect silos and build bridges between your entrepreneurial spirit and the pulsating heart of AI. Consider this a blueprint for your enterprise, a guiding light as we explore the fusion of timeless principles and cutting-edge technology. *We stand on the edge of a revolution*, and the time for small businesses to stake their claim is now.

In an era where technological milestones are seemingly reserved for the Silicon Valley elite, **a profound shift is unfolding—a renaissance where the power of Artificial Intelligence (AI) is no longer shackled to the monopoly of tech conglomerates**. This evolution in AI is shattering the illusion that small businesses are mere spectators in the digital revolution. **The truth is, advanced AI tools are now within arm's reach of the small business owner, beckoning a new era of efficiency and opportunity.** Just as David stood before Goliath, armed with little more than a slingshot, today's entrepreneurs are equipped with AI to challenge the giants of industry and win the battle.

The democratization of AI is not a tale of the future; it is the unfolding narrative of the present. **AI as a service (AIaaS) platforms have sprung up, offering plug-and-play solutions that can be seamlessly integrated into existing business frameworks.** No longer is there a need for colossal budgets to develop bespoke systems from the ground up. **The modern marketplace offers tailored AI tools that cater to the unique needs of small businesses—in customer service, sales, marketing, operations, and finance - all wrapped in pricing models that respect the limited resources of smaller entities.**

Entrepreneurs are wise to acquire knowledge of AI and discern its applications for their ventures. There exists a growing repository of case studies, tangible success stories illustrating how comparable small businesses have harnessed AI to streamline processes, predict market trends, and engage with customers more personally and efficiently.

One might consider the strategic application of AI as an act of good stewardship, an entrepreneurial embodiment of my favorite Biblical parable: the Parable of the Talents (Matthew 25:14-30). Utilizing AI in a small business context is akin to investing your talents, multiplying resources and capabilities, with a foresight that anticipates and adapts to shifting customer needs, creative destruction, and changing market dynamics. Success belongs to the business owner with the commendable management and growth of what one has been entrusted with, regardless of the initial scale. AI enables small businesses to magnify their impact, achieving more with less—turning modest operations into ecosystems of productivity and innovation.

Let us dispel the mistaken belief that some business owners have, maybe even you, that AI adoption necessitates a radical overhaul of one's business model. On the contrary, the integration of AI can be incremental, aligning with the philosophies of agility and lean operations. Starting with automating mundane tasks such as schedule management or customer inquiries through chatbots can free up invaluable time for business owners and their teams — time better spent on creativity, higher-level tasks, and growth strategy. These small yet significant steps fortify the foundations of a business, preparing it for sustainable scalability with AI as a steadfast ally.

The richness of AI for small businesses lies in its versatility. A local retailer, for example, can leverage AI to manage inventory more efficiently or quickly analyze customer sentiments through social media feedback. A small-scale manufacturer might implement machine learning algorithms to predict equipment failures and mitigate downtime while such equipment gets repaired or routinely maintained. The accessibility of AI today empowers such businesses to be proactive rather than reactive, to be creators of their destiny rather than bystanders or hostages to market forces outside of their control.

Unlocking AI's Potential: Transformation Across Business Functions

Evidently, the landscape of small business is being remodeled by AI, molding entrepreneurs not into followers but into leaders in innovation. The succeeding sections will delve deeper into the various arenas where AI can serve as a catalyst for transformation. Prepare to explore how the practical application of AI is not just about embracing new technology, but about cultivating a new mindset—a mindset that sees beyond the constraints of size and embraces the limitless potential of intelligence: artificial, divine, human, and beyond when combining more than one.

Embracing the Digital Artisan: AI as the Modern Small Business's Ally

In the realm of small business, weaving AI into the fabric of day-to-day operations can be likened to the skilled hands of an artisan enhancing their craftsmanship. Just as the artisan molds their inputs with finesse and vision, AI equips entrepreneurs with the tools to sculpt their business practices with unprecedented precision and creativity. For the discerning business owner, it's essential to recognize AI as not solely a technological tool, but as a means to harness one's ingenuity and multiply the fruits of human labor with faithful stewardship.

Scriptural teachings often remind us that it is through wisdom that one can build and establish. In line with this principle, AI serves as a repository of knowledge and understanding, helping turn vast data into actionable insights. It is the modern equivalent of the 'wise counselor' in business, guiding owners in decision-making and strategic planning. AI simplifies complex analytics, allowing small businesses to make informed decisions akin to those of much larger entities.

A Symphony of Efficiency: AI in Operations and Logistics

Consider the operations of any business, a collective orchestra playing the symphony of daily activity. **You, as the Chief Executive Human (CEH) can be the conductor, enabling harmony and efficiency in this complex performance powered by AI.** By automating routine tasks, it frees up valuable time, allowing entrepreneurs to focus on critical thinking, strategic planning and growth plan execution. This divine allocation of attention aligns with good and faithful stewardship, an essential concept within faith-oriented teachings that emphasizes the importance of managing resources wisely.

Automation is particularly transformative in the areas of inventory management and supply chain logistics. AI can predict fluctuations in demand or identify potential disruptions, prompting preemptive adjustments to help optimize daily operations. This proactive approach minimizes waste and maximizes profitability, embodying the scriptural directive to be prudent and faithful over the few you currently have, so you can be trusted with more afterwards.

Customer Interactions Transformed by Compassionate Algorithms

The spirit of service underpins every faith tradition and is a cornerstone in business. AI can elevate these customer interactions to new heights, showcasing an ethos of understanding and responsiveness. Chatbots and virtual assistants can offer personalized, considerate support at a national and global scale, reflective of the golden rule of treating others as one wishes to be treated. They provide responsive assistance at any hour, enhancing the customer experience without the constraints of traditional 9-to-5 operating hours, especially if you have customers in different time zones worldwide.

The granularity of personalization possible through AI also means small businesses can now offer a tailored experience that rivals that from larger competitors. By cultivating this intimate understanding and connection with customers, small businesses can nurture greater brand loyalty and establish a much larger community of advocates, turning customers into evangelists and raving fans — a true manifestation of creating deep, enduring relationships for the long-term.

Financial Prudence Augmented by AI Insight

Financial acuity is essential for any business, and in this area, AI is particularly transformative. The capability of AI to distill actionable and higher-level insights from vast financial data is significant beyond measure. I selfishly wish I had AI working with me when I was in finance listening to quarterly earnings calls, building financial models, and drafting investment recommendations.

Businesses are thus empowered to make financial decisions with greater confidence, equipped with more insights than a human team may not have been able to supply with the limited time they had on hand, ensuring that business owners, entrepreneurs, executives, managers and team leaders remain good and faithful stewards of their financial resources.

Whether it's through improved forecasting models, detecting irregularities in transactions, or optimizing pricing strategies, AI assists in navigating the financial landscape. This prudent management is not only about survival but also about thriving; it enables small business owners to invest more strategically in their mission and vision, now that they can dream a bigger dream than ever before as AI raises the bar on what they can achieve past their prior limitations.

Unlocking Innovation with Algorithmic Partners

As companions on the journey of innovation, AI algorithms offer the potential to break new ground in product and service development. They unearth patterns and opportunities that may go unnoticed by even the most experienced entrepreneur, revealing avenues for invention and re-invention. It is the proverbial lamp unto our feet, illuminating pathways we may not have traversed on our own, providing a testament to the endlessness of creation, innovation and re-imagination.

AI's role in Research and Development (R&D) is significantly altering the landscape of possibilities for small businesses. It allows them to undertake sophisticated research that would otherwise be prohibitive due to cost or expertise restraints. This capacity to innovate en masse democratizes the competitive landscape, allowing David to stand toe-to-toe with Goliath.

Sales and Marketing: AI's Strategic Flourish

In the back-and-forth dance of sales and marketing, AI plays a strategic partner, guiding small businesses in executing precise and impactful moves. Thanks to artificial intelligence, marketing campaigns can now be personally tailored, resonating more with the individual on the other end of the screen. AI's data-driven insights enable a customized approach that speaks directly to the customer's heart, mirroring the highest ideal of personal connection between customer and business.

Sales processes are also enhanced through AI's analytical prowess, identifying leads with a higher likelihood of conversion and facilitating more effective upselling strategies. These approaches are foundational in building a sustainable business, reflecting an approach that doesn't merely seek to sell but to serve the customer by offering genuine value in what he or she truly needs to maximize the overall experience with your product or service.

Unveiling the Path Ahead

As we progress through the narrative of integrating AI into the everyday fabric of small businesses, it is clear that this technology is much more than a tool — **AI is a collaborator that augments human intention** with its computational might. It can complement our strategies that align with the most profound tenets of spiritual wisdom: stewardship, service, and innovation. With each passing page, we shall explore these areas more deeply, unwrapping the layers that form a comprehensive blueprint for transformative action and potentially exponential growth in the sales, profits, and the potential exit value of the business. With the help of AI, the entrepreneurial spirit is not just sustained but uplifted, bearing witness to expansion, enrichment, and excellence in the realm of small businesses worldwide.

The Economic Epistle of AI

When we consider scripture, we are often reminded of the wisdom of stewardship and prudent resource management. *"Whoever can be trusted with very little can also be trusted with much,"* (Luke 16:10) could very well be a directive for how small businesses can wisely embrace AI. Herein lies an analogy with a seed – you wouldn't need a vast field to prove that a seed can grow; a small patch of fertile ground suffices to demonstrate its potential. Artificial Intelligence, in a similar vein, can germinate within a small enterprise, growing into a magnificent tool that significantly reduces operational costs and profitably increases its revenue.

It is through the implementation of AI that small businesses can automate repetitive tasks, such as customer inquiries, inventory management, and basic record-keeping, thereby freeing up human resources to focus on more strategic initiatives and higher-level tasks that require the creativity, critical thinking, and emotional intelligence that only a human can reliably provide.

The Bridge to Enhanced Customer Relations

AI transcends the boundary of mere cost-cutting and finds its noble purpose in enhancing customer experiences. Just like faith without works is dead, customer service that lacks personalization and promptness will fail to inspire brand loyalty and consequently risk driving customers to your competition. AI makes it possible to tailor interactions with customers based on their preferences and past behaviors, ensuring that each engagement reflects a deep understanding of their needs. For a small business, this translates into AI-driven chatbots, recommendation engines, and responsive support systems that operate with an almost divine insight into what the customer seeks.

The Gift of Informed Decisions Through AI

To deliberate and decide wisely, one must have the right information. In the case of small businesses, the use of AI for data analytics mimics the act of seeking knowledge and understanding. **AI has the unique ability to analyze large datasets quickly and accurately**, identifying trends, predicting customer behaviors, and optimizing pricing strategies. Equipped with these insights, a small business owner can pivot with precision, turning data into actionable knowledge.

The Virtue of Accessible Technology

What is so exciting about AI today is that it has become increasingly accessible to all who seek to employ it. This is a reflection of how cloud-based AI services have opened the gates of innovation, **allowing small businesses to subscribe to AI tools without the need to invest heavily in their own IT infrastructure**. These services, which range from customer service bots to analytics and forecasting tools, are often pay-as-you-go or subscription-based.

The Quiet Revolution in Back-Office Operations

In the stillness of a prayer room, one often finds peace inside and an order that the outside world lacks. Likewise, AI introduces a peace of sorts to the back-office operations of a small business. Through process automation, it establishes a realm of efficiency where business processes are handled with celestial accuracy and speed. Invoicing, payroll, and financial reporting that once consumed hours of human labor (ask me how I know firsthand!) are now managed in the background, faultlessly and quietly, freeing the human spirit for more creative and strategic endeavors that AI cannot easily replicate.

Unearthing Hidden Efficiencies

AI's role in small business mirrors the tale of the hidden treasure; within its algorithms and data patterns lie unseen efficiencies waiting to be uncovered. It is akin to the careful refinement of a raw gem – as AI is integrated into business systems, it illuminates ways to streamline operations, reduce waste, and improve service delivery. By adopting AI, businesses bring forth hidden efficiencies that may have remained buried under the mundanities of daily operations, including which "fire du jour" the business owner needs to put out today vs. yesterday vs. tomorrow, much like discerning individuals pursue a higher calling beyond the toils of day-to-day existence.

AI: The Harbinger of Future Prosperity

Let us not be mistaken; harnessing the power of AI is no casual endeavor, and it should be approached with the same reverence as one would handle a sacred text - with diligence, focus, and conscientiousness. AI does not replace human ingenuity; it augments it, providing tools that can breathe new life into traditional business models. AI, in its essence, is not merely a tool but a harbinger of transformation and prosperity for those with the wisdom to embrace its potential.

Unveiling the Power of AI for Your Business Venture

AI is the steward of the modern-day small business, nurturing efficiency and growth in ways that were once unimaginable for the small business owner years ago. The myth that AI is the exclusive realm of corporate giants has been dispelled; akin to David's five smooth stones, AI is the accessible tool that offers even the humblest of businesses a chance to meet their Goliaths head-on.

AI serves as a catalyst for transformation — a concept echoed in the parable of the talents. Just as the faithful servant invests his talents to multiply them, so too can businesses invest in AI to multiply their effectiveness across operations, from marketing to customer service. Empowering your business with AI is akin to the wise builder who constructs his house on a rock; it provides a **resilient foundation** in a swiftly evolving business and technology landscape globally.

The Journey Forward

This chapter is but the prologue to a grander narrative awaiting your enterprise. Within the remaining pages of this book, anticipate an odyssey illuminated by actionable advice, reflections from personal experience, and authoritative insights. Embrace this journey with the same vigor with which you pursue all worthy endeavors: passionately, persistently, and clothed in the whole armor of resilience and foresight.

Prepare for a revolution in your business mindset. Unlock the power of AI, step by indomitable step. Let anticipation rise, for you are about to embark on a transformative adventure that promises not just survival but thriving success in a world where the smart not only survive but soar. Trust in these truths, invest in this wisdom, and draw from the wellspring of opportunities that AI integration offers. Your business, no matter its size, can gain significant ground by wielding artificial intelligence with **strategic intent**.

The stage is set, the tools accessible, and success awaits you.

Are you ready to join the AI revolution?

2

—·—

CUTTING THROUGH THE COMPLEXITY: AI ADOPTION SIMPLIFIED

In the lambent glow of the early morning, Eli, a small-town grocer, stood before the modest array of produce in his store, hands buried deep in his apron pockets, contemplating the quiet landscape before him. The tight rows of apples and oranges were more than mere fruit; they represented a lattice of possibility, a symbol of the growth he envisioned for his business. The digital age had ushered in a paradigm shift, and Eli felt the call to evolve, but hesitation tugged at the hem of his ambition.

Eli's mind wove through the passages of Proverbs as he contemplated integrating artificial intelligence into his operations; *"Plans fail for lack of counsel, but with many advisers, they succeed."* How often had he leaned on these words for guidance through the years? Yet the lingering misconceptions about AI — its cost, complexity, and scale — whispered doubt into his dreams. Like David before Goliath, Eli sensed that confronting these giants with the sling of such new technology might pave the way to a victory unexpected by many in his neighborhood.

The chime of the door echoed through the quiet store as Mrs. Abigail, a regular, entered. The interaction with her was intimate, a testament to Eli's belief in personal connection. However, the efficiency AI promised could afford him the luxury of nurturing these relationships even further, a thought which heartened him. "Good morning, Eli," she greeted, her voice carrying the warmth of familiarity. His response was both a salutation and an affirmation of purpose.

Eli mulled over evangelists in various disciplines who professed that with the right tools, any David could tackle Goliath. Cloud computing, subscription-based models, the premiere of user-friendly AI — the promised land seemed ripe for harvest to those willing to take the first step. He pondered upon the economic principle of comparative advantage; just as nations thrive by focusing on their strengths, so must he leverage technology to enhance the personal touch that set his business apart from his competitors.

The melodies of a faith-driven life played like a gentle undercurrent throughout his day—Eli often found solace and strength in these heavenly rhythms. Pausing, he surveyed his domain: the potential lushness of digital integration, like the promised land, awaited his courage. "The strength of the diligent leadeth to abundance," he recalled from Proverbs 21:5, the wisdom of scripture washing over him, a balm for his trepidation.

As he stocked shelves, the simple, repetitive act took on the quality of meditation. In every apple placed, every sweep of the broom, there was an echo of the greater mechanics at work—the precision of systems, the call to efficiency, the dance of order and opportunity. In these moments, the complex became manageable, the tableau of a grander design, each decision a brushstroke upon the canvas of his future.

Eli's reflections were as seeded fields, ready to bring forth the fruit of action at harvest time. But the sowing requires more than contemplation; it requires faith and works. And as dusk crept across the sky, painting it with hues of resolve, the question lingered in the air, as palpable and electric as the shifting winds of change: Would Eli courageously embrace the journey before him, and lead his business into the renaissance of artificial intelligence?

Dispelling the Myths: AI for Every Small Business

The landscape of business competition is witnessing a seismic shift with the advent of artificial intelligence (AI)—a tool that, despite common misconceptions, is increasingly within reach for small businesses. Within the sacred texts, there is an understanding that wisdom is not the preserve of the mighty alone; the book of Proverbs reminds us that *"wisdom is calling out in the streets"* (Proverbs 1:20). In a parallel conviction, AI is no longer an oracle for the corporate elite but a practical tool beckoning small business owners. **It is time to address the widespread misunderstandings about AI's affordability and complexity, and shepherd entrepreneurs towards a future where AI integration is as fundamental as a solid business plan.**

Skepticism towards new technology is a tale as old as time, but the Scriptures have taught us that with knowledge, we can cast away fear. A business owner's trepidation in facing the perceived Goliath of AI technology can be overcome through the David-like sling of accurate information and guidance. By **realizing that AI solutions exist on a spectrum that accommodates the capacity and resources of small enterprises, we start to chip away at the colossal misapprehension that AI is out of reach for them.**

In the heart of this revelation lies the recognition of AI's versatility — offering scalable solutions that can be custom-tailored to a company's size, niche, and budget. Like the mustard seed in the parable (Matthew 13:31-32), these initial AI implementations may start small, but with the potential to grow into something substantial with proper nurturing to protect your business with much needed shade and stability. **The age of prohibitive costs and labyrinthine systems is yielding to a new era of user-friendly AI-powered platforms, often managed without the need for specialized teams or exorbitant external consultants.** This progression aligns seamlessly with the standards of good stewardship, where making the most of every resource is not just sound business practice, but a spiritual imperative.

To take full advantage of what AI can potentially offer to your business, a mindset shift is urgently required; the business world needs leaders that can see the horizon of change and act accordingly. This chapter aims to foster just that: a visionary approach for small business owners, grounded not in prophecy, but in practical actions and strategies for AI adoption. The integration of AI should be seen as a reverent act of preparing one's business for the future, akin to the way one would sow the land with trust in the harvest to come.

This chapter's mission is to equip small business owners with the mental frameworks and factual insights necessary to navigate the AI adoption journey confidently and effectively.

Practical AI Solutions: Thriving in the Small Business Arena

The tools of AI adoption are many and varied, ranging from intuitive off-the-shelf software to comprehensive service applications that can revolutionize customer service, marketing strategies, and operational efficiency.

Throughout this exploration, we will examine case studies of businesses that have successfully integrated AI to enhance their services and products, thereby reaping both financial and strategic rewards. As you immerse yourself in their stories, let their journeys instill in you the belief that what was once considered the domain of industry titans is now accessible to the industrious entrepreneur.

Busting Myths about AI and Your Business

Small businesses may shy away from AI adoption due to the fear of steep learning curves and intimidating technology. However, such concerns tend to obscure a vital truth: advancements in AI have paved the way for more intuitive interfaces and straightforward applications. In the spirit of the parable of the talents, there exists a responsibility to optimize the resources at hand, and AI is increasingly becoming one of those resources.

The barrier of perceived expense is another significant hurdle. The assumption lingers that AI solutions come with an exorbitant price tag—a luxury only affordable to well-established enterprises. Yet, in practice, many AI services operate on a subscription model that scales with usage, thereby ensuring that even the smallest of businesses can adopt AI without crippling their finances. In line with the principles of good stewardship, it becomes possible to judiciously allocate funds towards AI integration, elevating a business's capabilities without endangering its fiscal health.

Aside from cost, another prevalent myth is that AI solutions are too generalized to be useful for niche markets or specialized small business needs. On the contrary, the flexibility and adaptability of modern AI mean it can often be tailored to specific industry requirements. From a bakery automating its inventory management to a consultancy refining its customer relationship tools, AI is not a monolithic entity but a diverse array of tools that can adapt to a spectrum of entrepreneurial environments.

Moreover, **the perceived complexity of implementing AI** does not account for the support ecosystems that have arisen to help you understand and adopt AI. Rather than navigating an opaque forest of code or being on hold for 20 minutes over the phone to reach tech support, small business owners can access a growing community of AI service providers, online resources, and user-friendly platforms to support them along the way of AI adoption and integration. With these supports, the integration of AI becomes not an anxious solo expedition but a guided journey, with plenty of assistance available for those who seek it.

It is important to dispel **another fallacy: that AI is an all-or-nothing proposition requiring a total overhaul of existing systems**. Many AI tools are designed to integrate seamlessly with existing software, providing a harmonious blend of old and new. This integration is gentle on both the budget and the learning curve, allowing for incremental transformation rather than disruptive transitions. Rather than being overwhelmed with too much information or too many tasks to implement just to get started, AI integration can be a gentle stream of change that is manageable, accessible, and poised to yield fruitful outcomes.

From busting up these fearful myths about AI, small business owners are empowered to unlock the potential of AI without succumbing to fears of impracticality or financial strain. The stage is set for a strategic embrace of AI — one that is informed, intentional, and integrative.

Tailored AI Solutions Give Way to Opportunity

AI technology is now scalable, adaptable and designed to grow with your enterprise. They extend their capabilities as the demands of the business expand, ensuring that you can always harness the most relevant and beneficial features for your business.

Cost-effective AI Subscription Models

The availability of subscription-based AI services negates the myth of cost-prohibitive AI, allowing small business owners to access potent tools without significant upfront investment. These agile models can offer a host of services on a pay-as-you-go basis. You no longer need to invest in costly infrastructure or expansive teams — AI subscription services embody the mustard seed philosophy, starting small but with the potential to grow into something vast and impressive in service.

Democratizing AI for Non-technical Users

Gone are the days when one needed a team of engineers to implement and manage AI tools. Companies now offer AI systems with drag-and-drop, user-friendly interfaces, designed with the layperson in mind, ensuring straightforward, peaceful interactions with technology. Such systems can perform complex data analysis or automate routine tasks, all the while being something that even a non-technical user can command and utilize effectively. They foster a harmonious environment where the tool serves the visionary, and not the other way around.

Personalized Customer Experiences at Scale

Personalization at scale might seem like a daunting task for a small business, yet AI brings it within reach, casting the net wide while ensuring each customer feels uniquely attended to. Through AI-powered recommendation engines and customizable communication tools, small businesses can employ strategies to tailor experiences and build relationships to a deep enough level across such great breadth that were once the stronghold of larger companies.

A Partner in Growth and Innovation

Utilizing machine learning algorithms, a small business can enhance its product offerings, improve service delivery, or enter new markets, all with the scaled guidance of AI. This harmony of human ingenuity and artificial intelligence is echoed in Ecclesiastes 4:9 — *"Two are better than one, because they have a good return for their labor."* Together, the pairing of your entrepreneurial spirit and AI's analytical prowess can yield prosperous outcomes for your business and the people it serves.

Progress with Ethical Underpinnings

Finally, it's vital to approach AI integration with ethical considerations at the forefront, ensuring the technology is employed in a manner that aligns with core values and societal good. This mirrors the ethical teachings found across cultures and religions, urging us to treat others with respect and fairness. As the Bible says, *"A good name is more desirable than great riches; to be esteemed is better than silver or gold"* (Proverbs 22:1). So too should a business that includes AI, striving to maintain a reputation for integrity and benefit to the community. No new technology is worth quickly, severely and irreparably destroying your reputation. AI is only as ethical as the human that controls it.

AI as an Integral Part of Your Team

Consider AI not as a replacement for the human touch that character-izes small businesses but as a complement to it. **Envision AI tools as diligent team members, available 24/7, tirelessly analyzing data, offering insights, and automating the mundane — thus freeing your human workforce to engage in more creative, higher-level, and strategic endeavors.** By reframing AI as an extension of your team, a sense of partnership evolves that integrates technology with the personal service that customers cherish.

Cultivating a Culture of AI Literacy

Finally, fostering a culture that understands and embraces AI is akin to sowing seeds on fertile ground. **Encouraging a culture of con-tinual learning and curiosity ensures your team remains agile and equipped to harness AI's benefits as AI continues to evolve over the long term.** Your leadership in this realm can serve as a beacon, one that not only illuminates the path for your own enterprise but also casts a wider light for other small businesses to follow. In this shared endeavor, the collective wisdom gleaned from embracing AI innovation can uplift entire communities, binding together the threads of economic prosperity and shared human advancement.

As entrepreneurs, the invitation is to **lean into the future with optimism, cultivating a landscape where AI is an integral tool in realizing the dream vision upon which your business was founded.** This endeavor is not merely about technology; it's about the harmonious blending of human creativity with digital potential—a symphony of progress rooted in timeless principles and forward-thinking adaptation.

AI Integration Strategies That Work

Drawing upon personal experience, I can attest to the transformative power of embracing AI. A close colleague once faced the same crossroads, with reservations about the foreign nature of AI in her boutique marketing firm. However, through iterative learning and integration of a customer relationship management AI-powered tool, she witnessed a dramatic improvement in client satisfaction and loyalty. Let this serve as a testament to AI's potential to revitalize and enrich your small business, no matter how daunting it may initially seem.

You now can choose to have your business's narrative, after being introduced to AI, not be one of hesitance or fear but of courage and strategic foresight. Now that we have successfully debunked some of the most common myths and fears surrounding AI, as we move forward, carry forward this spirit of innovation and resilience, preparing to weave AI into the very DNA of your small business success story.

3

— · —

AI WITH A HEART: PRESERVING THE HUMAN TOUCH

I t was a crisp morning in the heart of Silicon Valley, where an unas-suming office building hummed with the fervor of quiet industry. Jonathan, a small business owner with a steadfast belief in the sanctity of personal touch in business, sat in a corner office that looked out over rays of golden sunlight. He contemplated the ethical dance of integrating artificial intelligence into his customer service model. His gaze wandered beyond the landscapes of code and algorithms, seeking the fine line where technology met humanity.

His hands rested on a worn leather-bound Bible on his desk, a gift from his grandfather, which often provided solace and wisdom. As he flipped through the Book of Proverbs, his thoughts swirled around re-taining the irreplaceable human element within his growing company. While the Proverbs spoke of wisdom and knowledge, Jonathan knew that applying them in this new digital age required a discerning heart. "How can we harness this AI as a shepherd uses his staff – guiding, yet not controlling the flock?" he pondered, looking at the screens of metrics and chatbots.

The click-clack of employees' keyboards was a testament to the creative work that his team delighted in. Jonathan had always held the belief that every individual should have the ability to bloom in their vocation, and AI, he reasoned, should be the gardener nurturing that growth, not a thunderstorm threatening to wash it away. Still, his mind wrestled with scenarios, weighing possible outcomes like Solomon, striving for an equilibrium of efficiency and ethics.

A knock on the door broke his reverie, and in stepped his lead engineer, Maria. She carried a calmness that seemed to stem from a soul anchored deep within her faith, a quality that Jonathan admired. "We've streamlined the support procedures with the new AI protocols," she said, "and there's more time for the team to innovate, they're genuinely excited."

Jonathan nodded, feeling a spark of hope. His vision had always been to empower his employees – to give them the freedom to create, to build relationships with customers that went beyond transactions. "We must ensure that this technology serves them and by extension, our clients, enabling our employees to sow seeds of genuine connection," he mused aloud, his resolve hardening.

As afternoon turned to dusk, and the office buzz settled into a quiet hum, Jonathan knew the delicate balance he sought was much like the constant turning of the Earth – a challenge yet a necessity. The golden shades faded from his office, whispering promises of a tomorrow filled with new possibilities and responsibilities.

Had he found a way to pave the path towards a future where technology and humanity might walk hand in hand peacefully and collaboratively? As dusk gave way to night, Jonathan sat at his desk, the silence enveloping him like a prayer shawl, and he considered: "Can we wield our advancing technology as David did his sling – a tool electric with potential but bound by a covenant to enhance and not to overthrow?"

The Harmony of Humans and AI in Small Business

In the ever-evolving landscape of small business operations, Artificial Intelligence (AI) emerges not as a usurper of human roles, but as an enhancer of the human touch that underpins ethical business practices. At the heart of this technological transformation is a profound and immutable truth: **technology serves best when it amplifies human strengths and virtues.** Indeed, ethical AI use, when rooted in spiritual wisdom, can complement rather than replace human roles, embodying the scriptural principle that all tools and inventions are to be used as extensions of God's creation, enabling humans to fulfill their potential and stewardship on Earth.

One can draw parallels between the harmonious integration of AI in small businesses and the biblical symbiosis of body and spirit. Just as the body serves the spirit, so can AI serve the human aspect of business, enabling entrepreneurs to tirelessly serve their customers with both efficiency and compassion. AI streamlines workflows in a way that can maintain, or even enhance, the personal rapport between a business and its customers – a testimonial to working smarter, not harder. Small businesses that adopt AI can balance the scales between transacting and relationship building, ensuring that no customer feels like a number in a ledger or a spreadsheet.

Empowering employees to focus on high-value, creative tasks is at the core of ethical AI deployment. This empowerment echoes the parable of the talents, where the servant is entrusted with resources and expected to use them wisely to create more value. In a similar fashion, AI acts as the trusted 'talent,' given to employees to multiply their effectiveness and creativity within their roles. With routine and repetitive tasks relegated to AI, employees have the space to pursue roles that require emotional intelligence, innovative problem-solving, and personalized service – the very attributes that define the human spirit at work.

From a real-world perspective, the implementation of AI in customer service provides a revealing case study. An AI that handles basic inquiries frees employees to resolve complex issues that require a nuanced and empathetic approach. Such an act is not merely a business decision; it is an affirmation of human value. **It signals a belief that the time and talents of employees are too valuable to be spent on tasks that machines can perform just as well, if not better, freeing employees to engage deeply with their customers and create meaningful connections and relationships.**

This ethical strategy reaps dividends not just in efficiency, but in fostering a workplace culture that values contribution, creativity, and community. The integration of AI should be envisioned as the construction of a cathedral, with each technological block placed thoughtfully to support the grand design of human-centric service. **AI with a heart** is the stone that supports the stained glass window – it holds firm so that the more delicate, artful work of human hands can be appreciated in all its colorful intricacy.

A Call to Humble Leadership

By implementing AI in ways that align with the highest spiritual and ethical standards, entrepreneurs not only preserve but endorse the human qualities that define their businesses. They also rise to a kind of humble leadership that acknowledges the role of the Divine in guiding technological advancement for the greater good of the communities they serve.

The pervading anxiety around Artificial Intelligence tends to revolve around **an age-old fear: the displacement of human jobs.**

Yet, this perspective does not consider the full spectrum of possibilities that AI presents. Ethical integration of AI into the workspace can accentuate and complement the unique capabilities humans bring to the table. In the pages of Holy Scripture, we find that each part of the body has a purpose, working in harmony to function as a whole. In a similar spirit, AI can become a seamless extension of the human workforce when approached with wisdom and discernment.

AI's capability to handle mundane, repetitive tasks is often seen as a double-edged sword. On the one side, there is efficiency and consistency; on the other, potential unemployment. However, when leveraged ethically, AI serves as the canvas upon which human creativity and strategic thinking paint a brighter future. **This symbiotic relationship allows for staff to explore roles requiring empathy, relationship building, creative problem solving, moral judgment, or complex decision making**.

One real-world example is the healthcare sector where AI assists with administrative tasks or diagnostics—creating spaces for healthcare professionals to engage more deeply in patient care and interpersonal connection. In the sanctity of such interactions, the human touch is not only preserved but is given a new lease of life thanks to the time recaptured by AI's assistance.

Small business owners embody a dynamic blend of pragmatism and vision, and AI complements this by freeing them from the shackles of operational tedium. Rather than spending hours on bookkeeping, inventory management or scheduling, AI tools can manage these tasks with precision, allowing entrepreneurs the freedom to strategize, innovate, and personally engage with their clientele. It is akin to having a trusted steward manage the day-to-day so that the business owner can focus on cultivating growth and nurturing customer relationships.

Empathy through Algorithms: The Human-AI Collaboration

We witness in real time the growing capabilities of AI to handle routine inquiries through chatbots and virtual assistants. These AI tools provide immediate responses to customer queries, which is vital in today's fast-paced world. Yet, the essence of maintaining a human touch lies not in the initial response but in the depth of the follow-up.

For instance, when a highly-personalized, AI-generated response leads to further conversation, we as humans step in to provide a nuanced, empathetic understanding that deeply connects with our clientele. This is where the AI-supported ecosystem really flourishes — machine precision coupled with human empathy creates a service experience that is both efficient and warmly personal.

The Personal Touch in Automation: The AI-Enhanced Human Role

Embedding AI into customer service processes enables team members to focus on complex issues that require a distinctly human touch. It's about elevating the role of human workers to craftsmen and craftswomen of customer experience, using tools and insights provided by AI to sculpt a more refined service encounter. As a result, team members develop a greater sense of fulfillment from their work, knowing they are providing service that truly resonates with the customer on a personal level. It is this symbiosis of human and machine that embodies the essence of ethical AI use — amplifying human potential rather than substituting it.

Transforming Challenges into Opportunities: The AI-Human Synergy for Problem-Solving

Consider the unconventional problems that often arrive on the doorstep of customer service—a delayed shipment, an unexpected product issue, or a unique customer request. AI can help identify patterns and suggest solutions based on vast datasets, yet it's the human employee who, guided by spiritual principles like patience and understanding, transforms these challenges into moments of connection and trust-building with the customer. AI, then, is the compass that directs us toward solutions, but it is our hands that carve the path and make the journey alongside our customers.

Crafting a Future Together: The Human-AI Partnership

As we forge ahead integrating AI into our businesses, it is crucial that we do so with intentionality and wisdom. AI is a powerful assistant in our quest to serve our customers better, but its true value is unlocked only when it is directed by human hand and heart—ensuring that our businesses not only thrive in efficiency but also flourish in the authenticity of personal connection. This partnership between human and AI is not just a convenience; it is a strategic alliance, one that reinforces the virtues of compassion, service, and ingenuity that underpin successful and ethical business practices.

By mindful implementation, AI can help us achieve a harmony where our client relationships are stronger, our business practices are smarter, and our workforce is empowered to bring forth their best daily — showcasing that AI, when directed by a caring heart and a righteous soul, isn't just smart business; it's also wise and compassionate.

Elevating Roles through Intelligent Assistance

In the office, AI-based tools can analyze data at an extraordinary speed and scale, transforming raw numbers into insightful reports. This rapid processing allows business leaders and employees to make informed decisions swiftly, ultimately **elevating their role to strategists and visionaries**.

Through freeing staff from the constraints of repetitive tasks, we also free their spirit to pursue professional and personal growth. AI presents a gateway to lifelong learning and enrichment, fostering an environment where skills are continuously sharpened and passions explored.

Harmonizing AI with Customer Service

AI's ability to handle routine interactions, such as scheduling appointments or answering FAQs, does not sever the lines of personal connection but strengthens them. By allowing AI to assume these tasks, **customer service representatives are afforded more time to resolve complex issues, provide bespoke advice, and build deeper relationships with customers**. Crafting an environment where empathy and interpersonal skills are valued above all else is in harmony with the virtuous pursuit of service to others.

Balancing Technological Advancements and Ethical Practice

In an era where artificial intelligence (AI) is reshaping the boundaries of human endeavor, it is easy to lose sight of ethical principles that have grounded us for ages. As entrepreneurs and custodians of this technology, we hold a sacred duty to ensure that our embrace of AI adheres to ethical guidelines and spiritual teachings. Proverbs 3:27 implores us not to withhold good from those to whom it is due when it is in our power to act.

A Call to Integrate with Integrity

As custodians of AI in the business world, we are called to merge this powerful technology with our heartfelt values. Let this be a call to action: to invest in AI with diligence and integrity, ensuring that each technological advance respects and uplifts human dignity. Let us be encouraged by the possibilities that ethical AI integration presents. As you navigate the intricacies of implementing AI in your operations, remember that mastery over this technology is not just a strategic asset but a testament to your commitment to living out your virtues while serving humanity. With this commitment, businesses will not only flourish economically but will also contribute to a richer, more compassionate world where technology benefits us all.

4

—.—

TAILORED TECH: CRAFTING YOUR AI STRATEGY

Mid-morning light shimmers through the tall windows of Lydia's quaint bakery, dust dancing like jubilant sprites between the sunbeams and the display of artfully crafted pastries. In her inner sanctum of creativity, she stands steadfast, a seasoned entrepreneur, rolling dough with the mastery of her craft and years of wisdom guiding her hands. Yet today, there is a weight upon her, a consideration that has been quietly disturbing the usual serenity of her early working hours.

She recalls the recent articles, the conversations, the incessant buzz about artificial intelligence—the modern loom that promises new patterns in the fabric of commerce. Her business, an emblem of personal touch and human sweat, would it stand to benefit from the charm of such technology? The musings interrupt her methodical movements. The prospects of AI—would they be an alchemic element to transform and transcend, or an unwelcome intruder in her realm of authenticity?

The fragrance of freshly baked bread is like an old friend to her senses, a reminder of tradition, yet part of her—perhaps the visionary entrepreneur in her—longs to taste the future, to delve into these uncharted waters. She envisions a system where AI could manage her inventory, predict the ebb and flow of patrons, perhaps even tailor treats to the burgeoning desires of her customers. Visions of risk and reward waltz in delicate balance as she ponders of the possibilities.

Lydia knows she stands at the crossroads of decision, a testament akin to the parables she had read, where the mustard seed of faith had the potential to uproot mountains. Her faith tells her that every tool laid before mankind holds a purpose under the heavens—a time to plant and a time to uproot. Could the advent of AI be the time to plant anew, for growth that could extend leaves and branches far beyond her imaginings?

It's in the intimate conversations with her loyal customers where doubts find their counterbalance. They speak not only of pastries and bread but of lives, dreams, and the very essence that weaves community. Here lies her business acumen, nurturing relationships just as she nurtures her dough. The strategy for AI integration must be deliberate, a golden thread woven into the tapestry of her business's unique design.

She envisions starting small, perhaps an AI program to personalize customer service, a taste before the full feast. As she presents a warm croissant to a waiting child, whose eyes light up in simple, pure joy, Lydia knows her journey will be one of careful steps, the rhythm of her resolve in sync with her dough kneader.

In the dance of flour and foresight, where does one find the courage to venture into such a delicate intermingling of tradition and innovation? How does an entrepreneur like Lydia weigh the scales of faith and business, merging the reverence of age-old wisdom with the bright pulse of her business's rising future with AI as its yeast?

Embarking on A Tailored Journey

Entrepreneurs set sail on a journey that, much like the story of Noah, requires them to anticipate what lies ahead and prepare meticulously for the voyage (Genesis 6:14-22). This journey involves recognizing the importance of a customized AI-enhanced business environment. Meticulously evaluating all aspects of the business to ascertain the most advantageous areas for AI integration is akin to understanding the depth of the waters before charting a course. Only then can one confidently navigate through the upcoming AI-fueled business landscape.

In this chapter, we will share methodologies and best practices that offer both the courage and the counsel to adopt AI in a way that maintains the essence and core values of your business. The focus is not on the adoption of technology for its own sake, but on facilitating growth, enhancing customer experience, and ensuring that AI adoption is both meaningful and sustainable.

As entrepreneurs, **navigating the integration of artificial intelligence (AI) into business operations is akin to tailoring a bespoke suit. It isn't just about choosing the fabric — AI — but about custom-fitting this technology to the unique contours of your enterprise.** The approach cannot be generic, for just as every person's measurements are different, so too are the needs, capabilities, and objectives of each business singular in their form.

Your AI strategy should be crafted to serve not only your business goals but also your higher purpose. It demands an approach that considers not just the tangible benefits but the profound impact AI can have on your customers, employees, and the broader community your business serves.

To tailor your AI integration effectively, you must first conduct a detailed analysis of your business operations, identifying areas where AI can be most beneficial. It is through this discernment that you can reap the optimal rewards of AI technology. For example, if customer service is the lifeblood of your business, an AI-enabled chatbot might revolutionize how you interact with clients. But if your business thrives on creative designs, then AI can be leveraged to streamline administrative duties, freeing up your human talent for more imaginative pursuits.

Each step forward in this tech-savvy landscape should be taken with calculated precision. **Pivoting to AI simply because it's a modern trend amounts to chasing the wind**. As Ecclesiastes 1:14 says, *"I saw all the deeds that are done under the sun; and see, all is vanity and a chasing after wind."* Instead, let your strategy be molded by the unique shape of your business challenges and aspirations, ensuring that the integration of AI serves a concrete and beneficial purpose.

Be not daunted by the complexity of new technology; instead, embrace it with clarity and confidence. Every decision on this journey should be purposeful and measured twice before cutting, for once integrated, AI becomes a thread in the fabric of your company's operations. Let your strategy emerge from a place of insight and adaptability, steadfast in the face of obstacles, and enlightened by the possibilities that lie ahead.

Evaluation: Assessing Your Business's AI Blueprint

As stewards of your ventures, honing the craft of discerning the true needs of your business is paramount when contemplating integrating AI. Begin by systematically identifying the areas where AI can significantly enhance operations. Ask yourself, "What tedious tasks could be automated? Which processes, if optimized, would exponentially bolster my business's service or product?"

Inventory your current challenges. AI's magic lies not in its universality, but in its customizability. Small business owners are often closest to the pulse of their operations and are uniquely positioned to pinpoint bottlenecks or inefficiencies that AI could address. Acknowledge that deploying AI is not about harnessing a trendy technology; it's about identifying and applying a solution that addresses concrete challenges—like reducing costs, speeding up production, or personalizing customer experiences.

Assess the potential benefits. Imagine the possibilities that could unfold from a thoughtfully integrated AI system: increased accuracy, faster response times, predictive analytics that enable better decision-making, or more time for your team to engage in creative endeavors. Let your vision be broad and aspirational, inviting the potential benefits to inform and motivate your strategy.

Weigh the strategic fit. Just because AI can be applied in many areas doesn't mean it should be applied in all areas of your business. It requires discernment and strategic focus. Consider the compatibility of AI with your strategic priorities. Does it align with your mission? In what ways? Will it truly serve your customers, or is it a departure from your core values?

Examine resource allocation. Implementing AI is an investment, and like any investment, it must be made with wisdom and foresight. Balance the costs against potential returns. Scrutinize both the initial and ongoing expenses of AI integration against the time and resources it frees up and the additional revenue it may generate for you.

Consider the impact on your team. Beyond analytical and financial considerations, AI can have profound implications for your workforce. As Proverbs 27:23 advises, *"Be sure to know the condition of* *your flocks, give careful attention to your herds."* Consider the skill sets, adaptability, and morale of your employees: how will AI integration affect their roles and responsibilities? Will it empower and upskill them, or could it cause disruption and anxiety? Develop a plan that leverages AI's capabilities while also valuing and developing your human talent.

Involve stakeholders in the decision-making. Consult with your team, customers, and possibly even your suppliers about possible AI applications. Acquiring diverse perspectives can provide invaluable insights and foster collective buy-in. Involving your stakeholders can harmonize various interests and reinforce a shared commitment to the business's growth and ethical use of technology.

Prioritize user experience. At the heart of any successful AI strategy is the user experience. If AI complicates rather than simplifies interactions or detracts from the personal touch that your business treasures, reconsider its deployment. As with all tools, AI should amplify — and not undermine — the value you offer. Hence, as you draft your AI roadmap, use the Golden Rule from Matthew 7:12 as a guide: *"Do to others what you would have them do to you."* Ensure AI elevates the experience for those you serve in the way you would appreciate it.

Gradual Integration: A Prudent Path

Prudence dictates a gradual approach, starting small and then scaling as you learn and as your business mandates. Initially, you might automate simple tasks such as customer inquiries with a chatbot or streamline appointment scheduling. These steps can offer immediate improvement in customer experience and operational efficiency, serving as tangible indicators that you are moving in the right direction. Such incremental changes allow you to measure impact, refine processes, and increase your AI fluency without overcommitting resources.

Comprehensive Restructuring: A Strategic Overhaul

For entrepreneurs inclined towards more dramatic transformation, a comprehensive restructuring to integrate AI can be the answer. This approach involves a systemic change where AI becomes a core component of all business operations. If your business intelligence suggests that an all-encompassing AI strategy aligns with your mission and that the resulting efficiencies will catapult you ahead of competitors, then such a bold move may be warranted. Undertake this with clear objectives and a robust support system so that the change is sustainable and places your business at the forefront of innovation.

Organizational Readiness: Laying the Groundwork

Before you embark on any form of AI integration, you should ensure that your organization is ready. This necessitates training your team, laying the technological groundwork, and creating a culture open to change. It is analogous to **preparing the soil before sowing the seeds**; nurturing your team's skills ensures they can adapt to and support the new AI-driven processes. This pre-emptive action mitigates resistance, fosters enthusiasm, and accelerates the adoption phase, thereby increasing the likelihood of your strategy's success.

Flexibility and Resilience: Core Tenets of AI Strategy

As you thread AI into the fabric of your business, **maintain a posture of flexibility and resilience**. The landscape of technology and business is ever-changing, and adhering too rigidly to a single strategy can lead to obsolescence. **You must be prepared to pivot, taking**

cues from both the successes and the shortcomings of your AI endeavors. Diversify your AI initiatives as you would in financial investments, exploring different applications and solutions to find the most fitting and impactful for your business.

Evaluating ROI: The Benchmark of Success

Financial prudence is key—evaluate the return on investment (ROI) of your AI integration. **This evaluation should probe deeper than merely tallying up costs against revenue; it should also consider customer satisfaction, brand reputation, employee morale, and long-term growth prospects.** Insightful analysis borne from robust metrics will lead you to make more informed decisions about your AI strategy and clarify the true value AI is adding to your business.

Ethical Considerations and Social Responsibility

As you strategize the deployment of AI, **uphold the highest ethical standards and consider the broader social impact of your decisions**. The integration of AI is not merely a business strategy but a commitment toward progress that respects the dignity and rights of individuals while fostering a society that benefits from technological advancements. Ensuring that AI is used in a way that aligns with moral principles and serves the greater good will enhance your brand's reputation and strengthen its legacy.

Looking Ahead: The AI-Empowered Future

As we progress in this narrative of AI integration, remember that this journey does not end with the implementation of a few tools or systems. It is about creating a future in which your business thrives amidst the waves of technological evolution. The strategies outlined serve as your compass, guiding you to craft an AI-infused enterprise where efficiency, innovation, and strategic foresight are not just ideals, but realities woven into the fabric of your business's everyday operations. Keep your gaze forward and your mindset attuned to the potential that AI has in transforming your business and the world it serves.

Conclusion

As we draw this chapter to a close, remember that **building a bridge to the future requires understanding the terrain of the present.** Aim to be well-versed in the language of technology, yet grounded in the time-tested truths of your faith and expertise. Always seek to build strategies that resonate with your business's core values and market position. In doing so, you will not only chart a course for success but also lead with purpose and integrity, lighting the way for others in the small business community.

5

GROW AS YOU GO: AI FOR EVERY STAGE OF YOUR JOURNEY

The chill in the air was rivaled only by the biting uncertainty in Jonah's heart. With each step through the sleeping streets of his modest town, his briefcase, heavy with contracts and business plans, seemed to mock him with its load—a testament to his growing enterprise, but also a burden of decision. His company, though once a sapling, had now stretched its branches wider than he'd ever dreamed, and the weight of scalability had wrapped itself around his waist like a shroud.

Through the cobblestone alleyways, tinged with the scent of rain-soaked earth, he pondered on scriptures that spoke of the mustard seed — a parable reflecting faith and growth from humble beginnings. Jonah's startup, much like the seed, had begun humbly, yet it now yearned for an infrastructure that could support its unforeseen expansion. Investing in adaptable AI systems was no longer a discussion for distant tomorrows; the future had arrived with the morning mist that clung to his tailored coat.

As he wound his way past shuttered shops, the warm glow from the bakery's windows caught his eye. The yeasty aroma of fresh bread called out, reminding him of simpler times, of routines unchallenged by success. Here was permanence, a continuity to aspire to, yet he also knew that without constant innovation, even the most cherished institutions could crumble, much like stale bread.

He passed by a mother tending to her garden, each flower an individual yet bound by a single patch of earth. Her nurturing hands guided Jonah's thoughts towards strategies that would allow his company to flourish without the need for frequent upheavals. The goal was stability in growth, adaptability in structure — akin to the garden's seasonal perseverance.

The sun peeked above the horizon, casting a golden hue on the slate rooftops, each ray a silent affirmation of the day's potential. A verse from Proverbs whispered in his mind, about committing one's work to the Lord, and trusting that one's plans will be established. Jonah clung to the notion that with foresight, his entrepreneurial endeavors could grow while maintaining their core, like a tree that stands firm regardless of how far its branches spread.

Much like the cycle of day and night, Jonah understood that his company's journey would go through phases. But unlike the predictable dance of celestial bodies, the path of his enterprise must be paved with strategic planning and powerful tools that evolve with time. Could it be that the right AI system would be the linchpin to achieving such harmony between growth and continuity?

Grow Wisely, Grow Well

Choosing the right AI tools can be akin to planting seeds in fertile ground. Just as seeds require space to sprout and grow, AI solutions should have the capacity to expand and evolve. For the small business owner, this means seeking out AI technologies that are not only effective at their current size but are also equipped to scale up as the business grows. Scalable AI ensures that as your operational needs become more complex, your tools can adapt rather than become obsolete, providing continuous support on your entrepreneurial path.

To identify scalable AI tools, it is crucial to start with an assessment of both current and future business needs. One should consider solutions that offer modular features or subscription-based models which enable incremental expansion. For instance, cloud-based customer relationship management (CRM) software can initially handle a modest client database but has the power to manage thousands more contacts over time with simple plan upgrades, without the need for a full system replacement.

Furthermore, embracing AI platforms with open API architecture can be a strategic move. This allows for seamless integrations with other applications and services as your business ecosystem grows. A point of sale system enabled with AI might initially help track sales and inventory; as you scale, its API could integrate with advanced analytics tools or e-commerce platforms, creating a more robust sales and operations engine that expands in capability alongside your business.

Equally important in a scalable AI is its ability to learn and adapt through machine learning. An AI system that becomes more efficient and smarter as it processes more of your data will continue to add value to the business. It should have the capacity to analyze large data sets and fine-tune its algorithms, improving decision-making processes, and unlocking new insights as business complexities multiply.

Investments in AI should also consider the longevity of the provider. Partnering with AI technology firms that have a track record of growth, innovation, and support ensures that they will continue to evolve their offerings in line with cutting-edge developments and growing business demands. Such partnerships can turn into long-term alliances that contribute significantly to the firm's sustainability and success.

On the human side of scalability, look for tools that offer user-friendly interfaces and lower learning curves. As roles within your company grow and change, new staff will need to interact with these systems. The AI solutions chosen should thus facilitate a smooth transition and quick adaptability for incoming talent, ensuring continuity without significant downtime or incremental training investment.

In the selection process, it is wise to engage with vendors and test out AI solutions through free trials or demos. This first-hand experience provides insights into the system's current performance and scalability potential. It also allows you to evaluate the provider's customer service, which is an important aspect of the business relationship as you will likely need their ongoing support to maximize the AI tool's potential as your company grows.

Adaptable AI systems are akin to a wise business associate who grows in knowledge and skill — only they do not seek advancement elsewhere. They enable cost-effective scaling and resource allocation, remaining a foundational aspect of your enterprise at any stage. As the AI grows more attuned to your business, the insights gleaned and efficiencies achieved become significantly more potent.

Strategic implementation of AI must go hand in hand with a continuous improvement mindset. This mirrors the spiritual idea of sanctification, where continual growth and perfecting of oneself is a lifelong process. Your AI systems should also continually improve, becoming more finely tuned to your business's unique patterns and challenges. This not only prepares your business for now but for the myriad of tomorrows that lie ahead, allowing you to face them with confidence.

To neglect the dimension of scalability and flexibility in your AI investments is akin to building a house upon the sand. It may stand for a while, but when the rains come down, and the floods come up, the house on the sand will fall with a great crash (Matthew 7:26-27). Thus, business owners should lay the foundations of their enterprises on AI systems that can expand and reinforce themselves against the impending tides of change and competition. This not only provides security but also a strong foundation for future growth.

Ultimately, the narrative of growth is continuous, and the tools you choose should exemplify that story. The value of adaptable AI is not just in what it does for you today but in its capacity to transform alongside your business. Thus, entrepreneurs are called not just to adapt to the present but to anticipate and shape their futures—one wise decision at a time.

In mapping out your company's future, proactively anticipate where your business may be years down the line. Just as Joseph stored grain in preparation for years of famine, as told in the Book of Genesis, you too must store ample technological capability for future expansion. This can be achieved by choosing AI systems with modular designs and open architectures. These systems allow you to add functionalities or integrate with more advanced technologies as your needs evolve.

Warning: at times, short-sighted decisions, such as opting for a cheaper AI solution, might seem attractive. However, it is the equivalent of building a house on a cracking foundation. Choose quality and flexibility over temporary savings to ensure that your business does not face system obsolescence too soon.

Proactive Maintenance and Incremental Upgrades

Maintaining your AI systems is no less significant than cultivating your spiritual life—both require continuous attention and care. Establish regular maintenance protocols to keep your AI systems healthy and updated. Schedule incremental upgrades to steadily introduce new features and capabilities. Start with small, manageable updates that cumulatively lead to significant improvements over time, thereby avoiding the tumult of a full system revamp.

The Role of Partnerships in Sustaining AI Ecosystems

In the pursuit of a scalable AI infrastructure, collaborative relationships with tech providers are invaluable. Such partnerships can pave the way for easy access to upgrades and support, ensuring that your AI system is not only current but also future-proof. Forge alliances with those who share your vision of growth, and together, create a synergy that will fuel your entrepreneurial endeavors.

Embrace Change, but Maintain Your Core Values

Although change is inevitable and often beneficial, it is essential to anchor your business in its core values. As your company and its AI systems mature, stay true to the mission and vision that define your enterprise. In times of transformation, these values will be your North Star, guiding your decisions and ensuring that growth is not just an increase in size, but an enhancement of character and impact.

Embedding Flexibility in Business Processes

Lastly, to harness the full potential of scalable AI, your business processes must also be flexible enough to accommodate change. This means creating workflows that are not overly dependent on specific technologies or platforms. It is akin to constructing a vessel able to navigate both calm waters and raging storms—resilient, adaptable, and purpose-driven.

Embrace these strategies, commit to adaptable systems, and invest in the future-ready AI tools that will empower your venture. Use the insights garnered from theology, business acumen, and personal experience to guide you, and together with these AI solutions, write the next chapter of your success story.

6

— • —

EMPOWERING MINDS: THE KEY TO AI FLUENCY

In the small but bustling office of a burgeoning tech startup, Julia stood by the window, sunlight cascading across her face, as the city's symphony floated up from the streets below. Her team, a mosaic of earnest ambition and raw talent, clustered around a whiteboard – a canvas of dreams sketched in dry erase marker.

Julia's mind replayed the sermon she'd listened to only the Sunday prior, where the preacher had spoken of the parable of the talents, an urging to be stewards of what we're given, to grow and multiply our gifts. In this case, her gift was her business, and she saw AI as the fertile soil to plant the seeds of progress.

There was Tom, who had stayed up all night poring over online courses on machine learning, and Mira, who had found inspiration in a recent seminar about natural language processing. Julia took pride in the culture she had cultivated, one of perpetual learning, a greenhouse for minds eager to embrace the transformative power of artificial intelligence.

This was not mere work to her; it was a quest for knowledge, guided by a belief that through understanding came power. In her view, AI was not just an avenue for financial gain, but a tool to unlock human potential, to empower her employees to reach for heights unknown.

The clock on the wall ticked, a metronome keeping pace with the heartbeat of progress. Julia's thoughts wandered to the countless small businesses teetering on the brink of transformation, if only they knew the steps to take. She pondered creating a simple guide, merging the principles of her faith with the pragmatism of her work – a beacon to illuminate the path through the fog of technological uncertainty.

As the afternoon waned and the enthusiasm of her team crescendoed into plans and prototypes, Julia knew the task ahead was mammoth, yet not insurmountable. She found solace in the scripture of Philippians 4:13, *"I can do all things through Him who strengthens me,"* for she believed fervently that with her team's collective strength and the grace of providence, they would not just adapt, but thrive in this new era.

The question that lingered, weighty yet hopeful, as the sun dipped behind the skyline and cast shadows across the room, was not if they would surmount the challenges of AI adoption, but what new horizons would they uncover when they did?

Harnessing the Power of AI Starts in the Mind

When small business owners confront the frontier of artificial intelligence, they stand at the crossroads of innovation and tradition. For the entrepreneur eager to grow and compete, embracing AI technology is not merely a strategic move but a testament to their commitment to progress and adaptability. Knowledge and education in AI are the seminal roots from which a more proficient, efficient, and competitive business can grow. They transform bewildering code into a robust tool, breathing life into the aspirations of businesses and enabling them to punch above their weight in a crowded marketplace.

In this revolutionary tide, **small businesses that fail to ride the AI wave risk being swept under by competitors who've learned to surf these digital swells**. Therefore, the priority is not just to adopt AI, but to garner a profound understanding of its capabilities and applications. The essence of technological empowerment is realized through a commitment to perpetual learning, a tenet that mirrors spiritual principles where the pursuit of knowledge is often hailed as a never-ending journey, a path to empowerment and enlightenment.

AI integration demands more than technical prowess; it requires an organizational culture steeped in curiosity, adaptability, and a sense of shared vision for the collective good. To forge such an atmosphere, small businesses should build a learning ecosystem where team members support each other in mastering AI tools and techniques.

The digital era we live in has democratized access to more information and education. Resources for AI education are plentiful from free online courses and webinars to seminars and workshops that demystify AI without the presumption of advanced technical expertise. These resources serve not only to enlighten but to empower teams, ensuring that every member—from intern to manager—has a foundational grasp of how AI can be harnessed across various business functions.

Navigating these resources is t**he first practical step towards achieving AI fluency**, providing teams with the insights and competencies necessary to harness AI for unprecedented benefits in customer satisfaction, operational efficiency, and innovative potential.

It is in this spirit that **entrepreneurs must encourage their teams to embrace a posture of lifelong learning**, with a clear awareness that wisdom and understanding are not static but evolve with exposure to new challenges and technologies. The dynamic interplay between human ingenuity and machine intelligence opens up a realm of possibilities as vast as the entrepreneurs' vision to grow their dream businesses.

A Blueprint for Technological Harmony: AI-Powered Customer Support

Step 1: Addressing the Echoes of Customer Concerns

Begin the journey of enhancing your customer support by listening—truly listening—to the voices of your clientele. These echoes of satisfaction or distress should guide your path. Delve into customer complaints, inquiries, and feedback; only by pinpointing the pervasive pain points can you offer a potential remedy that heals rather than merely numbs the symptoms.

Step 2: Surveying the Technological Landscape

With a clear map of customer grievances in hand, turn your gaze to what is available across the technological landscape to resolve them. Which AI-driven solutions resonate with the issues at hand? Be it a chatbot for instantaneity, a recommendation engine for customization, or a knowledge base for information accessibility, identify the right tools that promise deliverance from your business's current inefficiencies.

Step 3: Judicious Tech Adoption

Now, employ discernment. Evaluate potential solutions, considering cost, compatibility, and complexity. Seek demos and / or free trials, wield them as a craftsman would, ensuring your choice fits like an extension of your business ethos. Only solutions proven through testing should be woven into the fabric of your ongoing operations.

Step 4: Tailoring the Tool

In the customization of your AI solution, work with your chosen provider or summon the expertise from within your ranks. Configure the system with a craftsman's touch, ensuring that every cog and wheel aligns with your business's unique demands and adjust as necessary. Whether it's a conversational flow for your chatbot or a data feed for your recommendation engine, meticulous detail is key.

Step 5: Empowering the Human Element

Your customer support team is the driving force behind the wheel of this new tool, and their adoption and proficiency determine how smooth the ride will be for both them and your customers. Educate and empower them, ensuring that they are adept in conducting this new symphony of human-AI collaboration.

Step 6: The Cycle of Refinement

The AI system is but a seed; it requires careful nurturing and husbandry. Monitor key performance indicators, always ready to prune and optimize. This is the garden of customer satisfaction, and it requires an attentive gardener to flourish through the seasons, while also rooting out any weeds that interfere with your growth.

Step 7: The Art of Customer Communication

Finally, address your customers with transparency about the new ally involved in their service experiences. Highlight the enhancements, help ease the transition and welcome their feedback. After all, this metamorphosis in service is for them – their contentment is the true indicator if this tool adoption is a success or not.

In this simple step-by-step approach to integrating AI tools into your customer support, the clarity of process and intention can lead to a harmonious blend of technological efficiency and human warmth. As such, the steps serve not only as a structure to be followed but as a dance to be performed - one that evolves with the music of changing customer needs, evolving technology and growing business goals.

To be fluent in the language of AI is not merely to code or build algorithms; it is to understand its potential impacts, ethical considerations, and its transformative role in commerce. Practical knowledge translates into realistic assessments of which processes can be automated or enhanced by AI, a step that could mean the difference between flourishing and floundering in the current economic landscape.

A most crucial aspect of this educational pursuit is the **awareness of AI's limitations and ethical implications**. As stewards of technology, it is incumbent upon entrepreneurs to operate with a sense of responsibility, having the discernment to know when and how to apply AI in a manner that is beneficial and not detrimental to society at large.

Discover the Pathways to AI Mastery

Let us then explore together the pathways to such mastery. Where can small business owners find the resources to propel their AI education forward? How can they instill a culture that not only embraces AI but evolves with it? The next segment ventures into the practical answers, outlining clear strategies for AI fluency that are integral to remaining competitive and visionary in an ever-changing digital econosphere.

As entrepreneurs at the forefront of the business battlefield, your arsenal must be as versatile as the challenges you face. **Enriching your team's understanding of artificial intelligence stands as one of the most formidable tools at your disposal**. Let's navigate the vast landscape of educational resources designed to bolster the AI fluency of your team.

Harness the Power of MOOCs

Online courses, particularly **Massive Open Online Courses (MOOCs)**, offer a gold mine of knowledge for those seeking to grasp the intricacies of AI. Platforms such as Coursera, edX, and Udacity provide courses from top universities and institutions, exposing your team to foundational theories, the latest tools, and real-world applications. Leverage these resources to build a structured learning path tailored to your business needs. Whether it's through a series designed by industry leaders or an academic syllabus, these courses can transform your team into adept navigators of AI terrain.

Dive into Digital Libraries and E-Learning Hubs

In the spirit of continuous learning and growth, digital libraries and e-learning hubs emerge as beacons of wisdom. Websites like Khan Academy or MIT OpenCourseWare offer free access to a treasure trove of knowledge. IEEE's AI and Ethics in Design course weaves vital moral considerations into the technical narrative, mirroring the Proverbs' wisdom in providing guidance for conduct. In these repositories, journeys of education are as diverse as the books of the Bible — each piece of content, a chapter leading your team closer to enlightenment in the age of artificial intelligence.

AI Podcasts, Webinars and Newsletters

Podcasts and webinars speak to the heart of modern learning—active, on-the-go, and engaging. Encourage your team to tap into podcasts like "The AI in Business Podcast" or attend webinars hosted by AI experts. By the way, **you can also search and subscribe to my *Use AI to Grow Your Business* newsletter on LinkedIn**. These mediums not only provide insights into emerging trends but also instigate stimulating conversations, much like the dialogues of the Socratic method, fostering a communal quest for understanding within your enterprise.

Encourage Engagement with AI Communities

The realm of social learning beckons with its promise of collective wisdom. Offer your team the opportunity to engage with AI communities on Reddit, LinkedIn, or Stack Overflow. These forums create an environment where seasoned professionals and novices alike can share experiences and solve problems together.

Utilize The Tech Giants' Educational Platforms

Tech giants, the modern-day Goliaths of innovation, extend slingshots in the form of their very own educational initiatives. Google's AI Education, Microsoft Learn, and IBM Skills Gateway present specialized courses that empower small business teams to utilize AI tools effectively. It's an invitation to stand shoulder-to-shoulder with these industry titans, wielding their resources to carve out your niche in your competitive market.

Expand Horizons with AI Books and Literature

Books on AI, such as "Life 3.0" by Max Tegmark or "AI Superpowers" by Kai-Fu Lee, serve as silent mentors to learn from via reading. Consider arming your team with such texts (**including another book of mine entitled *Future-Proof: How to Adopt and Master Artificial Intelligence (A.I.) to Secure Your Job and Career***) to deepen their understanding and inspire a forward-thinking mindset that anticipates the evolution of AI in enterprise, as well as their careers.

Foster Mentorship and Workshop Opportunities

Enlightenment through personal interaction can surpass the pure transfer of knowledge. Seek out mentorship programs and workshops led by AI practitioners who have walked the path your business is now embarking on. **If I can be of help towards growing your business with AI, contact me and learn more on how we can potentially**

work together on my website (DreamBusinessMakeover.com).
In these collaborative settings, the Proverbial principle of "iron sharp-
ening iron" comes to life as participants exchange hard-earned nuggets
of practical wisdom, honing their collective skill sets and expertise
together.

It is not simply the acquisition of knowledge that elevates your
team, but the manner in which it is harnessed, applied, and shared
that truly defines the journey. Arm your team with these resources and
watch as they grow from learners to leaders in the realm of artificial
intelligence, each step illuminated by the light of education and guided
by a relentless pursuit of innovation.

An Organizational Culture Rooted in Curiosity and Adaptability

As small business leaders, one of the most profound gifts we can offer
our team is the encouragement to remain perpetually inquisitive — a
quality that aligns closely with the scriptural call to seek wisdom continuously. In the dynamic landscape of artificial intelligence, instilling
a culture of curiosity and adaptability isn't just a choice; it becomes
a sacred duty. **It's about fostering an environment where questions are welcomed and exploration is not just permitted but
expected**. Practically, this might materialize as brainstorming sessions
focused on how AI can address business challenges, or it might involve
empowering team members with time and resources to research new
AI technologies.

Modeling the Way

All endeavors to cultivate a particular culture must begin at the top. **As chief stewards of our businesses, it is incumbent upon us to model the behavior we wish our teams to emulate**. In the Bible, Paul's letter to the Corinthians about imitating him as he imitates Christ can serve as a powerful metaphor for leadership in business. Our actions ought to reflect an ongoing commitment to learning and growth, especially when it comes to AI. By staying abreast of AI advancements and openly discussing these with our teams, we demonstrate that knowledge and adaptation are vital for the health and growth of our enterprise and thus is a growing part of our company-wide culture.

Continuous Learning as a Core Value

Incorporating continuous learning into the core values of your company enshrines its importance within the organization. It also acts as a pledge, reminding us of the fluid nature of knowledge, especially within the realm of artificial intelligence. By creating a strategic plan that includes AI education as a fundamental objective, we encourage our team to dedicate themselves to perpetual learning, which in turn, helps them to skillfully navigate and leverage AI advancements for our business.

Leveraging Diverse Learning Opportunities

Capitalizing on the diverse range of learning opportunities available today can greatly augment the AI fluency of a team. A multi-faceted approach that combines online courses, in-person workshops, and periodic training sessions can **address different learning styles and preferences, ensuring no team member is left behind**. Affording team members access to prominent AI conferences or inviting guest speakers from academia or industry can spark inspiration and broaden understanding of how AI is evolving.

Encouraging Experimentation and Embracing Failure

A pivotal aspect of fostering a culture of adaptability is to **embrace experimentation — and inevitably, failure**. As innovators in the business community often understand, failure is not a permanent state but a stepping stone to greater understanding and eventual success. This is especially true in the application of AI, where trial and error can lead to profound insights and breakthroughs. By **creating a safe space for experimentation, where failure is seen as a valuable part of the learning process**, you encourage team members to test out new AI tools and methods without fear of retaliation or embarrassment.

Rewarding Innovation and AI Advocacy

To further solidify an ethos that places a premium on adaptation and AI integration, **consider establishing recognition and reward systems. Acknowledging team members who excel in implementing AI solutions or who are active proponents of AI education can greatly incentivize others**. It creates an environment where striving for excellence in AI literacy and application becomes a shared goal, celebrated by all.

Integrating AI Ethics in Everyday Discourse

Lastly, infuse discussions of AI with a strong ethical foundation, understanding that **with the power of these technologies comes great responsibility**. Drawing on ethical teachings from religious texts and philosophical works can provide a strong moral compass that guides the application of AI in business. **Making ethics a part of regular conversation surrounding AI projects** ensures that decisions made will not only be technologically sound but morally grounded as well.

By encouraging a spirit of inquiry, a readiness to adapt, and a commitment to continuous learning, entrepreneurs unlock the full spectrum of possibilities that AI offers. This cultural transformation within the organization lays a foundation not only for the successful integration of AI but also for the creation of a team that is resilient, knowledgeable, and prepared for the future.

Empowerment through Knowledge

Proverbs 4:7 imparts a timeless piece of wisdom: *"The beginning of wisdom is this: Get wisdom. Though it cost all you have, get understanding."* In our journey with artificial intelligence, this call to seek understanding resonates strongly. The investment in knowledge and continuing education around AI paves the way for small businesses to not only adopt new technologies but also to thrive in an ever-evolving digital landscape. **Education is the cornerstone upon which we build a future-proof enterprise, capable of weathering changes and seizing opportunities within the realm of AI.**

Cultivating a Culture of Curiosity and Adaptability

Fostering a culture that embraces curiosity and adaptability transforms your team into proactive participants in the growth and transformation of your business. Your culture should emphasize agility, enabling it to implement AI applications with grace and confidence. **Encourage questions, reward initiative, and celebrate the diversity of thought.** This way, facing the complexities of AI turns into a collective pursuit of growth and discovery, progressing towards AI mastery for everyone.

Harnessing AI: A Faith-Driven Approach

The Bible teaches us that through wisdom, a house is built, and by understanding, it is established (Proverbs 24:3). By anchoring AI adoption in faith-driven wisdom and understanding, not only does one build a robust business but also creates a legacy that reflects ethical stewardship, loving service to one another, and fruitful innovation.

So let us step forth, armed with knowledge, guided by continuous learning, and inspired by a workforce cultivated to inquire and evolve. Trust in God's wisdom to light our path, and let our learnings from AI be not merely for profit but for the betterment of our communities and service to His creation. Rise to the occasion with determination, and remember, the power to harness AI for your small business resides not solely in the machines but also in the empowered minds of your people.

7

— • —

PLUG-AND-PLAY PROGRESS: TAPPING INTO PRE-BUILT AI

In a modest office adorned with the relics of entrepreneurial battles, Simon logged into the CRM system, his fingers tapping with the hesitant rhythm of a man at the crossroads of innovation and the familiar. The gentle hum of his computer was a steadfast companion in the otherwise silent room. The glow of the early morning sun seeped through the blinds, casting stripes of light across the room that seemed to highlight the line between the traditional methods he knew and the burgeoning possibilities of artificial intelligence.

His business, once bustling with manual processes, now seemed on the brink of a transformation. Simon had built his company from the ground up, with sweat and conviction as mortar. Yet, in Scripture, there was a wisdom that spoke of new wine in old wineskins, and Simon pondered whether his reluctance to embrace change was akin to holding onto the old skins that would ultimately burst.

A new email notification from his business partner broke his concentration. "Check out the latest AI feature updates. It could be the breakthrough we need." The words both excited and unnerved him. He had seen large corporations deploy AI to great effect, but the world of small business felt to him like a delicate ecosystem, too fragile for such seismic shifts.

He recalled a recent bookkeeping error that had cost him dearly, the kind of mistake that AI could have caught instantly. A blend of Proverbs and pragmatism whispered to him; if wisdom could be found in the counsel of many, perhaps there was a place for artificial intelligence as an advisor in the multiplicity of his business decisions.

Simon knew that his company was a mosaic of needs, and the AI functionalities embedded in the tools he already used could streamline his operations. Maybe there was a divine nuance in the way AI could predict customer behavior or optimize his supply chain. Could the efficiencies promised be considered as manna from heaven for his contemporary business wilderness?

He saw his reflection on the computer screen—a face lined with the history of every deal and decision—and wondered if the time had come to step out on faith and embrace AI adoption. His old dogma clashed with new possibilities.

In the days to come, Simon would explore these AI functions, test their limits, pray for guidance, and perhaps activate features that had long laid dormant, as hidden talents waiting to be invested and grown to full potential. His hands hovered over the keyboard—was this the moment to part the digital seas and lead his business through untouched paths towards the promised land that would provide business's own version of milk and honey to him, his employees, and his customers?

As the morning light became bolder, carving deeper shadows across his desk, a question settled in his heart: could the leap into AI be a real-world application and testament to faith, the substance of things hoped for, and the evidence of things not seen?

Start with What You Have and Grow from There

There exists a misconception that AI integration necessitates starting from the ground up, building intricate models and systems tailored to specific business needs. Yet in truth, **many small businesses already possess the key to a seamless AI integration without even realizing it**. Prefabricated AI tools embedded within existing software can simplify the adoption process, making it an accessible reality rather than an elusive future prospect.

Consider the potential lying dormant within the software suites and applications you use daily. They are not merely tools for tasks but allies armed with latent AI capabilities ready to be harnessed. It becomes evident that the integration of AI need not be a daunting overhaul but a strategic awakening of capabilities already at one's fingertips.

Forging a path toward AI prowess begins with recognizing the resources already laid before you. This approach aligns with a deep-seated tradition of resourcefulness born from the wisdom echoed in spiritual texts. As the biblical parable of the talents teaches us, success comes not from the abundance of resources but from the stewardship of what we have been entrusted with. Small businesses should consider their current software as seeds of potential; with cultivation — activating AI functionalities — they can flourish into fruitful trees of efficiency and insight.

A business owner who approaches AI implementation as an evolution rather than a revolution enables a gradual and measured integration without disrupting the existing operational flow. It places the small business owner in the role of an adept craftsman who utilizes every tool with purpose and precision.

Unlocking Your Software's Hidden Potential

Action is the keystone of progress. To animate the dormant AI within your systems, one must first understand what features are available. This can often be uncovered through a simple exploration of your software settings or a conversation with your service provider. Once you are aware of these features, the pathway to activation becomes clear. Training sessions, webinars, or tutorials may be required to harness the full power of these AI capabilities. Yet, with each step, your business transforms, aligning closer to a vision that intertwines with the ultimate fulfillment of your enterprise.

The parable of the good steward should resonate now, not just as a distant principle but as a lived experience. When we wisely invest our "talents," which in our case might be the latent AI capabilities within our tools, we are rewarded with growth and efficiency. The text "to whom much is given, much is required" (Luke 12:48) underlines the responsibility of leveraging these advanced tools for the good of our customers and, by extension, our business and community.

Understanding the Alignment of AI-Enabled Tools With Business Growth

In the world of business there is a constant quest for efficiency. This call to diligent stewardship mirrors the proactive steps that small business owners can take by harnessing AI-embedded tools to bolster their operations. CRM systems, for example, are no longer simply repositories of customer contact details. By integrating AI, these platforms can now predict customer behaviors, **suggesting when a client might be ready to make a new purchase**, thus giving the sales force an incredibly powerful arrow in their quiver.

For example, small businesses in the bookkeeping industry have witnessed similar transformative innovations. Once tedious, the process of reconciling invoices can now be expedited through the use of AI, which not only speeds up the workflow but also minimizes human error. In this case, AI acts as **a tireless and precise counselor**, ensuring the financial bedrock of a business remains solid.

Leveraging AI for Customer Service Excellence

It is clear that in business, the customer is king (or queen), and exceptional customer service is the cornerstone of success. Reflecting on the notion that *"it is more blessed to give than to receive"* (Acts 20:35), we can appreciate the value of exceptional customer service, an area where AI is serving small businesses with grace and efficiency.

Chatbots, for example, enabled with AI, have the capability to not only address customer queries round the clock but to also learn from these interactions, continuously improving their performance. This **constant availability and adaptability are modern-day miracles** for small enterprises aiming to maintain a high level of customer satisfaction without the overhead of a 24/7 human workforce.

Enhancing Marketing With Data-Driven AI Insights

AI is equally revolutionary when implemented into marketing strategies, enabling businesses to **analyze vast quantities of data to identify trends and tailor campaigns that resonate deeply with their target audience**. These AI-derived insights could be seen as a form of wisdom, providing a pleasant bounty to a company's outreach efforts, ensuring that the message is not just heard but truly felt by those meant to receive it.

Streamlining Operations Through Predictive Analytics

Small business operations have been streamlined by predictive analytics, a form of AI that forecasts inventory needs, identifies potential logistical setbacks and aids in decision-making processes both big and small. Imagine the advantage to a business that, empowered by AI, can **anticipate and mitigate risks before they ever come to fruition**.

Empowering Human Resources With AI-Enhanced Recruiting

In human resources, recruitment processes are being elevated through AI-powered tools that can **sift through countless applications to identify the best potential candidates**. By applying this depth of insight, small businesses can more effectively discern the right individuals to join their team, **significantly reducing the time and resources often wasted on the recruitment process**.

Transforming Decision-Making With Business Intelligence Software

Finally, business intelligence software equipped with AI not only **sifts through mounds of data but also provides actionable insights and recommendations, which can elevate decision-making from guesswork to a strategic and informed process**. Here, AI provides understanding generously, assisting business leaders to navigate the complexities of their daily decisions as well as their monthly, quarterly and annual goals, budgets and objectives.

Check What You Already Have Available

For small business owners, the world of artificial intelligence can seem like a vast expanse of complexities, but in reality, harnessing AI's power is often as simple as switching on a light. Embedded AI features in popular business software sit, ready and waiting, like dormant seeds in fertile soil, needing only the refreshing waters of activation to sprout forth productivity and growth. To do this, one must often look no further than the settings or feature list of their current platforms. **Begin with an exploration of the platform's dashboard; seek out sections labeled 'Analytics,' 'Automation,' or 'AI.' These will often lead you to a treasure trove of AI functionalities at your fingertips.** Apply this wisdom and seek the concealed power within your tools to enhance your business' capabilities.

The Wisdom in Upgrading

Advancements in AI move at an agile pace, much like the changing of seasons. Ensuring your software is up-to-date is crucial—as it opens doors to the latest AI improvements. Upgrading may not always be free, but consider it an investment, akin to planting a vineyard that will yield fruit in season. Ecclesiastes 11:6 advises, *"Sow your seed in the morning, and at evening let not your hands be idle, for you do not know which will succeed, whether this or that, or whether both will do equally well."* In a similar vein, diversifying your technological investments by upgrading can lead to unexpected success in due season.

Utilizing Available Tutorials and Resources

Navigating new features can be daunting, but most platforms offer a wealth of tutorials and guides. Look within the Help section or community forums of the software for assistance. Harness these resources, and don't hesitate to seek live support when available. Don't shy away from asking questions; it is through inquiry that understanding is born.

Integration and Compatibility Checks

Before diving headfirst into AI features, it's imperative to ensure they are compatible with your existing systems and processes. Like the thoughtful builder who counts the cost before laying a foundation, conduct a thorough examination of how these AI integrations will interact with your business's digital infrastructure. Luke 14:28 teaches us, *"Suppose one of you wants to build a tower. Won't you first sit down and estimate the cost to see if you have enough money to complete it?"* Likewise, take stock of whether your current setup is ready to support the added AI capabilities you wish to activate.

Data Preparation and Organization

AI thrives on data—the more organized and clean your data, the better AI can work for you. **Evaluate the state of your business data to ensure accuracy and organization.** This task is reminiscent of tending a garden; removing the weeds and organizing the plants allows for a more robust and vibrant growth. Proverbs 27:23 encourages diligence in knowing the condition of your flocks — take this to heart and apply it to the digital assets of your enterprise that you shepherd.

Setting Clear Objectives and KPIs

After activating AI features, setting clear objectives and key performance indicators (KPIs) is key for measuring their impact. They serve as the milestones of progress on your business journey. Articulate precisely what you wish to achieve; whether it is improved customer service response times, increased sales through personalized recommendations, or reduced time spent on repetitive tasks. Let wisdom guide you in crafting objectives that will establish your business on a solid foundation of AI-enhanced operations.

Fine-tuning and Optimization

Post activation, the process of fine-tuning AI functions begins. Monitor the performance, gather feedback, and be prepared to make adjustments. This iterative process is much like tilling the soil — necessary to cultivate the optimal conditions for growth. It's also important to **remember that AI learns over time**; initial patience can lead to greater long-term effectiveness.

Embracing a Culture of Continuous Learning

Finally, a culture of continuous learning and adaptation is crucial when integrating AI into your small business. As AI evolves, so too should your strategies for its deployment. In this fast-paced digital age, **standing still equates to falling behind**; constantly educate yourself and your team on new developments in AI technology.

By following this practical guide to activating AI features within existing platforms, small business owners can prepare their businesses to reap the many rewards of this remarkable technology, ensuring their enterprise not only survives but thrives in the competitive landscapes ahead.

8

CHOOSE YOUR COMPANIONS WISELY: PARTNERING WITH THE RIGHT AI PROVIDERS

A midst the gentle hum of machinery, Michael surveyed his small factory floor. Wood shavings caught glints of sunlight streaming through high-set windows, carrying with them the scent of pine and the promise of productivity. His hands, still rough from years' worth of craftsmanship, now rested upon blueprints for the company's next big project. Yet something new was transpiring within the walls of his workshop, something that hummed with the potential of change.

His mind traced the edges of a decision yet made; the integration of artificial intelligence into his small business. "AI..." he mused, allowing the notion to mingle with his thoughts of sawdust and stain. "Could such a partnership truly understand the intricacies of a craftsman's needs?" His heart held a silent prayer for answers, hoping for a sign that could align his blue-collar company's values with the capabilities of modern technology.

The phone call earlier that day resonated within him, the voice on the other end spoke of tailored solutions, of support forged in empathy – an AI service provider that specializes in guiding small businesses like his own. It was a concept deeply rooted in understanding, one that didn't merely seek to sell a product but to walk alongside those it aimed to help.

Thus, with a careful hand, he penned thoughts on parchment, his considerations deeply etched with the wisdom of the ages, comparing and contrasting against the sayings of Solomon, the keen understanding of necessity and abundance. "There is a time for every purpose under heaven," he recited quietly, weighing the benefits, the risks, the leap of faith that stood before him.

His interactions with the environment were rhythmic, a patterned response to the natural ebb and flow of his routine. Yet, within this familiarity, his ruminations on forming a successful partnership with an AI provider were like new patterns emerging in well-known wood grains. It required a keen eye, a discerning heart, and the boldness to carve out new paths.

Would this partnership nurture his company's growth, or would it lead him into the impersonal automation he had always feared? What wondrous new creations could come forth from such an alliance? The juxtaposition of the old and new, the analog and digital, seemed like the merging of bygone eras with chapters yet to be written.

As the day neared its end, Michael stood by the open doorway, the cool evening air whispering hints of possibility. His eyes rested on the horizon, where the setting sun painted the sky in hues of deep orange and red – a daily testament to the world's infinite possibilities.

Could this potential union signify the same for his business – a horizon rich with progress and growth, yet still firmly rooted in the craftsmanship that had been its foundation since day one? And as a man of faith, did he not believe in embracing the tools provided for the prosperity of his work? Was partnering with an AI service that understood the soul of a small business the grace he had been seeking and praying for?

Forge your Path with the Right Allies

Selecting the right AI provider is a profound step in your small business's journey toward growth and sustainability. This partnership involves trust, commitment, and a shared vision for the future. When considering artificial intelligence to revolutionize your enterprise, it's not just about adopting new technology; it's about embracing a partner that comprehends the intricacies of your struggles and triumphs. A dedicated AI provider does more than supply technology; they become an extension of your business, providing guidance and support as you navigate the digital transformation.

In a world teeming with grand-scale AI solutions tailored for industry giants, **small businesses require a provider focused on the unique needs and finite resources of smaller operations**. To invest in an AI provider solely constructed for the herculean enterprises is to navigate a path of unnecessary hardship. Instead, seek a provider familiar with David's sling; minimal yet precise tools that amplify your tactical advantage against Goliath-sized competitors. This approach offers a more personal level of support, ensuring the provider understands the subtleties of your market and your company's distinctive fingerprint on the world.

Finding the optimal AI partnership demands vigilance and wisdom. **You must identify service providers that offer the empathy and customization synonymous with a bespoke artisan.** Much like discerning buyers search for those who craft goods with heart and intention, entrepreneurs should look for AI partners with a similar resonance. It is essential to probe potential AI partners on their understanding of small business demands, evaluating their responses and solutions like a skillful merchant assesses quality goods.

Once a suitable AI provider has been identified, the next step is establishing a partnership that encourages smooth integration of AI into your business processes. **Think of it as building a bridge**—each side must meet in the middle, supported by pillars of strong communication, shared objectives, and mutual respect. It should be a symbiotic relationship, one where both parties can thrive and learn from each other. The AI service provider should be an ally who not only provides technology but also imparts wisdom, much like a mentor guides a protégé.

Navigating the AI transformation journey requires not just tools, but also fortitude. Consider the early voyagers, who set sail into uncharted waters, relying firmly on the stars above and their seasoned navigational skills. Likewise, with the combination of AI and expert guidance, you can chart a course toward business breakthroughs and mastery of the digital realm. **Pioneers, at heart, embrace the change rather than cower from it**, and so should you, with the right AI provider by your side.

Finally, remember that adopting AI is a continuous journey of improvement and learning. Just as a gardener tends to their plants, nurturing them to bear fruit, **your partnership with an AI provider requires both patience and effort. It's through consistent communication, feedback, and iteration that the true potential of your AI solutions will unfold.** Your business will not merely survive but will thrive, propelled by intelligent systems that learn and evolve with you.

Nurturing the Vision Together

In crafting this partnership, see it as not just a contract between two parties, but as a covenant that binds your aspirations to the expertise of your chosen AI ally. With such a partnership, you can look forward to a future where technology is not a fearsome behemoth but a powerful steed you ride towards success. Your business shall not walk alone; for with the right companions, the path to a flourishing enterprise is walked in stride, imbued with innovative spirit and a steadfast resolve.

In Proverbs 13:20, there is a verse that extols the virtue of prudent partnership: *"Whoever walks with the wise becomes wise, but the companion of fools will suffer harm."* In the modern context of business and technology, this ancient wisdom resonates with particular clarity. For small business owners, choosing an AI service provider is akin to selecting a business ally — one whose expertise could either bolster your business endeavors or lead you into costly missteps.

Tailored Solutions: A Small Business's Ally

By gravitating towards AI service providers that place a premium on small business needs, there is an implicit understanding that your business's unique challenges and constraints are being seriously considered. Unlike larger corporations that might offer a one-size-fits-all solution, **providers focusing on small entities tend to design AI tools and services that are adaptable, affordable, and aligned with your specific business processes**. Such alignment translates to a quicker integration and better ROI, highlighting the tactical advantage of a tailored and nimble approach.

Unparalleled Support: The Empathetic Edge

Additionally, these specialized providers tend to offer more dedicated customer service and support. They empathize with the trials and triumphs of small enterprise — and why wouldn't they? After all, many of these providers may have started as small businesses themselves, navigating the same waters you find yourself in now. **Their support is both a service and a testament to their own journeys; therefore, they are vested in your success**, often providing a level of hands-on assistance and advice that larger, more detached providers simply can't match.

Navigating Complexity: Making AI Accessible

For entrepreneurs whose days are already stretched thin by manifold responsibilities, the idea of adding "AI integration" to the mix might seem daunting. Yet, **the right AI partner diminishes this complexity, distilling the sophisticated world of artificial intelligence into manageable, actionable steps** — transforming what might appear as an insurmountable challenge into a series of *achievable milestones*.

Scaling With You: Growth-Focused Partnerships

Imagine a partner who not only provides a service but also grows with you, evolving their offerings as your business scales. AI providers committed to small businesses are designed to do just that. Their solutions are built with scalability in mind, giving you the confidence that today's investments will continue to bear fruit even as your needs expand. This foresight is grounded in the provider's understanding that **your growth signifies their growth — a perfect synergy that can propel both companies forward.**

Empowerment Through Education: A Learning Curve Made Easier

Implementing AI isn't merely a technical upgrade; it's also an opportunity for education and empowerment. **Good provider partnerships offer continual learning resources, leading not just to the use of new tools, but to a stronger comprehension of how AI can be leveraged to innovative ends**. As you become more knowledgeable, you are better able to discern other areas where AI can be beneficial, fostering a culture of innovation and resourcefulness within your team.

Cost-Effectiveness: Maximizing Value for Money

In the economics of small business, budgetary considerations take precedence. **AI providers that prioritize small businesses are aware of this fiscal sensitivity and are attuned to providing cost-effective solutions that don't sacrifice quality.** By investing in such a partnership, you're ensuring that every dollar spent is working effectively towards enhancing your business's capabilities, optimizing your expenses and boosting your bottom line.

Shared Values: Cultivating Ethical AI Practices

Partnering with an AI provider that shares your values — especially when it comes to ethics and responsible AI practice — is critical. Companies that serve smaller businesses often hold these values in high regard, as they deeply understand the impact technology can have on people and communities. By aligning with a provider that holds ethical considerations at the core of their business, you ensure that the AI integrated into your operations will contribute to a legacy of integrity and trustworthiness, which is invaluable for customer relations and brand reputation.

Assessing Alignments: The Intersection of Vision and Technology

When selecting an AI service provider for your small business, it is crucial to seek out those whose vision resonates with the values and long-term goals of your enterprise. **An AI provider should not merely offer tools; they need to offer a visionary approach that complements your company's planned trajectory**. It becomes essential to **choose a partner that understands your business ethos and is genuinely invested in your success**. This shared vision is the foundation of an enduring and fruitful partnership.

Judging Expertise Through Demonstrable Success

Providers that boast a track record of success with similar-sized businesses give reassurance of their expertise. They have traversed the path you are embarking upon and know well the potential pitfalls and high points. Evidence of successfully implemented projects in your industry or comparable fields serves as a testament to their ability to deliver. **Look for case studies or testimonials, which often portray a clearer picture of what to expect.** Such tangible evidence not only speaks to the effectiveness of their solutions but also to their familiarity with the challenges specific to your business and industry.

Individualized Support: The Heart of Small Business AI Integration

One of the hallmarks of exceptional AI service providers is their willingness to offer tailored support. Unlike larger corporations, small businesses require more hands-on guidance and a provider that appreciates the nuances of more modest operations. They should offer customization options within their solutions — suggesting they perceive your business as unique and are thus more equipped to cater to your specific needs. **Providers with a reputation for crafting bespoke strategies are more likely to assist you effectively in integrating AI into your operations**.

Empathy in Action: Fostering a Culture of Understanding

Empathy goes a long way in any relationship, especially when it pertains to technological integration. **Select providers that demonstrate a genuine empathetic understanding of your position as a small business owner—ones that exhibit patience and take the time to comprehend your concerns**. A business steeped in compassion will likely offer more amiable and accommodating customer service, creating an environment where you feel comfortable discussing issues and asking for help. This is critical, as the implementation process will undoubtedly require ongoing support and collaboration.

Transparent Communication: The Key to Trust

Effective communication is pivotal; it builds trust and sets the stage for clear expectations. **Providers should be straightforward about the capabilities of their AI offerings and transparent about pricing and any additional costs. This clarity enables you to judge the return on investment unequivocally and avoid any surprises along the way.** Providers that communicate openly about the measures they take to ensure data security and privacy also underscore their commitment to your business's welfare in the digital realm.

Leveraging Knowledge: Continuous Learning as a Service

As a small business owner, your time is scarce. Thus, providers that offer comprehensive learning resources and training empower you and your staff to better leverage AI tools. **Look for those who go beyond the sale**, offering webinars, tutorials, or even personalized training sessions tailored to different levels of technical proficiency. This educational support not only underscores their dedication to your success but also facilitates smoother adoption by demystifying AI.

Flexibility and Scalability: Preparing for Growth

Finally, as your business evolves, your AI needs will too. **Opt for providers that emphasize scalability and flexibility—ones prepared to grow with you. Their services should be adaptable, able to expand and adjust in response to your business's changing demands.** This foresight is paramount, as it prevents the need for provider changes down the line, which can be costly and time-consuming. Establishing a partnership with a flexible AI provider ensures that, as your business flourishes, your AI capabilities can scale in tandem.

Forge relationships with AI providers that mirror the commitment and passion you invest in your own business. These partnerships, grounded in a shared vision, empathy, and a collaborative spirit, can dramatically streamline the process of AI integration. By doing so, you position your enterprise on a steadfast course towards a future where artificial intelligence is harmoniously woven into the fabric of your small business, yielding innovation and competitive advantage.

Aligning Spiritual and Business Values for Strong AI Partnerships

Entering an alliance with an AI service provider is not merely a commercial transaction; it reflects a profound alignment of values and vision. **Choose a partner that not only brings technical expertise but one that also appreciates the sweat and prayer poured into your small business.** This harmony can fortify your venture, turning technological integration into a cornerstone of your business's growth.

Establishing Clear Communication Channels

A successful partnership thrives on clear and consistent communication. King Solomon once said that life and death are in the power of the tongue (Proverbs 18:21), and in the context of business partnerships, this wisdom carries significant weight. Miscommunication can derail even the most promising collaborations. When establishing ties with an AI provider, ensure that there are defined channels and protocols for communication. **Transparency about goals, milestones, and expectations** will cement understanding and facilitate the seamless integration of AI into your operations.

Committing to a Shared Journey

Embarking on the path of AI integration is a commitment to a journey of growth and adaptation—not unlike the dedication to a spiritual path. As you seek to forge a partnership, **look for a provider willing to walk alongside you, one that is patient and willing to teach, much as a disciple walks with a mentor.** Remember, this is not a short sprint but a marathon, requiring endurance and mutual support. When challenges arise, and they will, a committed partner will provide guidance analogous to the proverbial lamp unto your feet (Psalm 119:105), illuminating the road ahead and ensuring you do not stumble.

Embracing Ethical AI Practices

Your chosen AI provider should not only enhance your business operations but should also adhere to the highest ethical standards. In the spirit of doing unto others as you would have them do unto you (Luke 6:31), **select a partner that prioritizes ethical algorithms and transparent practices**. The AI community is increasingly aware of the importance of ethical AI, and your business should be a beacon of integrity in its usage — not just for the benefit of your enterprise, but as an example to the industry and your customers.

Understanding the Financial Commitment

Be wise stewards of your finances. Just as Scripture teaches the importance of faithful stewardship and effective resource management (Matthew 25:21), ensure that the AI solutions you embrace are within your business's means. **Establish a clear understanding of cost structures and scalability options to avoid financial strain**. A provider sympathetic to the budgetary constraints of small businesses will be instrumental in finding the right balance between innovation and affordability.

Building Toward the Future Together

Finally, **look beyond the immediate horizon and evaluate the potential for long-term collaboration. A provider that not only solves today's problems but anticipates tomorrow's opportunities can help propel your business forward**. Envision a partnership that grows and evolves with your enterprise, reminiscent of the parable of the mustard seed which, though the smallest of seeds, grows into a large tree (Matthew 13:31-32). A shared commitment to weather setbacks and celebrate successes throughout the journey together underpins a partnership grounded in mutual respect and shared objectives.The right AI service provider is not just a vendor; they are an ally for the future, enabling your business to expand its branches and bear fruit and shade for years to come.

Through these principles, your partnership with an AI service provider will not only be commercially fruitful but rich in mutual respect and shared values. As you stand at this technological threshold, consider these insights not as the end but as the beginning of a transformative chapter in your enterprise's story. The right AI integration, underpinned by a strong partnership, can elevate your small business to new heights of innovation and success.

The Foundation of Successful Integration

Establishing a successful AI integration for your business is less an event and more a **cultivated relationship**. Smooth transitions are crafted through continuous dialogue, patience, and a commitment to adaptability. Just as **Ecclesiastes 4:9** empowers us with the insight that *"Two are better than one, because they have a good return for their labor,"* so too does a well-matched partnership between your business and the right AI provider yield abundant fruit.

Strive not for a mere transaction, but for a **synergetic alliance** that will navigate through the waves of technological innovation and marketplace evolution, allowing your business to soar on the wings of informed and intuitive AI application. These collaborations, bridging the human with the artificial, are the lifeblood of a business's thriving ecosystem.

Empowerment through Action

Let us take inspiration from the parable of the talents (Matthew 25:14-30) and remember that to whom much is given, much will be expected. Your decision to embrace AI is an investment—a talent entrusted to you with the expectation of growth. It is your responsibility, your mission, to partner wisely, to foster this gift with care, and to cultivate its potential to bear fruit in abundance.

Take bold action and steadfastly pursue the right partnerships that will underpin your small business's transformation. Adopt an ironclad resolve, and remain unwavering in your search for providers that resonate with your core values and understand the essence of your mission.

Enter the marketplace with confidence, harnessing the power of AI as you weave these modern threads into the fabric of your traditional business values. It is through these confluences of past and present, of human insight and machine intelligence, that we craft a future abundant in both prosperity and purpose.

Let us now move forward, keenly aware that the journey ahead is one rich with opportunities — a horizon expanded by the collaboration of human ingenuity and the marvels of AI. Find your path, enrich your enterprise, and let the partnerships you forge today light the way to a brighter day tomorrow.

9

COMPETITIVE CLOUT: AI IN THE GLOBAL
VIRTUAL MARKETPLACE

I n the dim glow of the early morning, Jacob stood amidst the quiet
kingdom of his small bookstore, the scent of old paper, and the
subtle creak of wood underfoot weaving a familiar comfort. His fingers trailed over the spines of books as if in silent prayer, an invocation
for wisdom. The world outside was awakening, but within these walls,
a different kind of renaissance was unfolding.

His mind wandered through the tales of prophets and poets, entrepreneurs of the spirit who spoke of mustard seeds and mountains moved by faith. For Jacob, his bookstore was not just a place
of commerce but a vessel for enlightenment. Yet enlightenment alone
does not balance books. In his heart, a burdensome question weighed
heavily: How to bridge the chasm between the hallowed past and the
fast-paced digitized future?

Artificial intelligence (AI) — the modern-day loaves and fishes,
multiplying resources where scarcity once reigned; a tool not unlike
the plow and the printing press in eras past. Jacob envisioned the
promise of AI as it could personalize his outreach to each curious
soul that sought refuge in literature's embrace, each recommendation
tailored as if divine by an unseen hand.

It was a Tuesday when he met with Nora, a young entrepreneur whose business in handcrafted jewelry thrived under the watchful gaze of algorithms and data patterns. She spoke of the global market with a fervor that resonated deep within him. *"With AI, one click unlocks the hearts of millions"*, she said. Her words danced in Jacob's mind, a hymn to the potential he yearned to harness.

As he nestled into the heart of his store, amidst the tomes of ancient wisdom and modern musings, he saw it clear. He would become a shepherd to his flock, guiding them to the books they sought, using technology not as a crutch, but as a staff to support and uplift. Data would lay out the map, and from it, he could discern the right paths, optimize what once was unimaginable — the whispered desires of a reader's heart.

He looked out upon the empty chairs where evening gatherings would spring to life, vibrant with discussion and debate, sewn together by the collective yearning for understanding. It was here that he would sow the seeds of a new beginning, a tapestry woven with strands of tradition and the golden threads of innovation.

As he contemplated his next move, the serendipitous chime of the doorbell ushered in the soft murmur of the morning's first customer, a testament to enduring traditions amidst change. And therein lay the crux of Jacob's revelation — could the wisdom of ages past illuminate the path for the small entrepreneur to flourish in a global market increasingly powered by artificial intelligence and machine learning?

Harness the Power of AI and Level the Playing Field

The advent of Artificial Intelligence has sparked a revolution, a beacon of hope for the Davids in a world of Goliaths. This technological renaissance has offered small businesses a mighty slingshot to race head-to-head with the industry giants. Once limited to local horizons, these enterprises can now forge their marks on the global stage, thanks to the prowess of AI-driven tools. What seemed like insurmountable walls are now gateways to a world of opportunities where innovation, efficiency, and scalability coalesce to turn the tides of commerce in favor of the small and agile.

Harnessing AI to Carve Your Niche: Tailor Your Message, Expand Your Reach

AI-Driven Personalization: A Competitive Edge

In the spirited terrain of the global virtual marketplace, personalization stands out as a beacon for small businesses seeking distinction. Artificial Intelligence (AI) unlocks the potential to tailor experiences to individual customer needs and preferences, a practice once reserved for the titans of industry, now within reach for the agile entrepreneur. Scripture reminds us to *"know well the condition of your flocks, and give attention to your herds,"* (Proverbs 27:23) suggesting the timeless importance of individual attention. Similarly, AI-driven personalized marketing mirrors this scriptural wisdom by **enabling business owners to understand and cater to each customer with a shepherd's care**. Such attentiveness can manifest as customized recommendations, dynamic email campaigns, and a uniquely tailored online shopping experience, creating a sense of belonging and appreciation among customers.

AI-Powered Customer Service: Serving With Excellence

Customer service stands as the cornerstone of business success — a truth as ancient as commerce itself. AI enhances this domain through chatbots and virtual assistants, adeptly handling inquiries with a precision that resonates with Proverbs' call for a wise and discerning heart. These intelligent systems are not only efficient but **can also operate in a multitude of languages, broadening the reach to a diverse global audience. With AI, businesses can offer impeccable service round the clock, transcending time zones and geographic barriers**, demonstrating an unyielding commitment to customer satisfaction.

Leveraging AI for Market Research: The Art of Listening

Before speaking to the heart of the customer, one must listen to their voice. AI excels in this area, collecting and analyzing vast amounts of data from social media, search trends, and online behaviors to glean insights into consumer desires and industry trajectories. AI-driven market research enables businesses to **craft messages that resonate with their audience's lived experiences and values**. By harnessing this information, entrepreneurs can design products and campaigns with a precision that feels divinely inspired, meeting the client's needs before they've even articulated them.

The Power of Predictive Analytics: Foreseeing and Acting

Anticipating the future is a skill highly praised in the books of wisdom — and with AI, small businesses can now approach this ideal. Predictive analytics allows for prognosticating sales trends, customer behaviors, and even potential supply chain disruptions. With this foresight, entrepreneurs can make strides in inventory management, marketing efforts, and strategic planning, **always remaining a step ahead**. As the sage navigates by the stars, so does the AI-equipped business navigate by patterns in data, steering clear of pitfalls and capitalizing on forthcoming opportunities.

AI-Enabled Price Optimization: Strategic Alignment of Value and Worth

In the delicate balance of commerce, setting the right price is both an art and a science. AI empowers businesses with price optimization tools that analyze market dynamics, demand fluctuations, and competitive pricing. This technology ensures that products are neither undervalued nor priced out of the market. By striking this equilibrium, small businesses can enhance profitability while maintaining fairness and stewardship that honors both their mission and their customers.

Streamlining Operations: The AI Advantage in Efficiency

Laboring wisely and efficiently is a principle that resonates through spiritual teachings and business practices alike. AI provides small business owners with operational automation tools that streamline tasks such as scheduling, inventory management, and invoicing. This newfound efficiency frees up precious time and resources, allowing entrepreneurs to devote their energies to growth and innovation — endeavors that reflect our inherent desire to create and improve.

Crafting an AI-Inclusive Strategy: Sowing Seeds for Growth

The journey toward integrating AI is akin to the careful planning and tending of a seedling. It begins with a clear vision and understanding of the unique challenges and strengths of one's business, coupled with the wisdom to adopt AI solutions that align with these qualities. As the Book of Ecclesiastes teaches, there is a time for everything, and **the time for small businesses to embrace AI is undoubtedly now**. With approachable tools and a strategic mindset, even the most modest of enterprises can flourish in the global virtual marketplace, harnessing AI to foster connections, innovate offerings, and stand out amidst the competition.

In the intricate tapestry of the modern global virtual marketplace, data is the golden thread that weaves through the fabric of business success. As entrepreneurs, recognizing that data-driven decision-making is not solely the purview of giant corporations can transform your small business into a formidable competitor.

With AI, analytics become democratized, offering insights that once required a team of data scientists. Now, a well-curated AI system can churn through vast amounts of data, uncovering patterns and opportunities that lead to informed decisions.

Efficient Inventory Management with Predictive Analytics

Predictive analytics powered by AI is a steadfast ally in efficient inventory management. This futuristic foresight allows you to stock appropriately, reducing waste, and optimizing cash flow. It's a practical application of the principle of stewardship, where every resource is used effectively for growth and sustainability. Implement AI-driven inventory algorithms to predict demand fluctuations and always be prepared to meet your customers' needs without excess.

Streamlining the Sales Process with AI Integration

The fusion of AI into your sales process injects efficiency and precision, akin to a master craftsman using the finest tools to create a masterpiece. **From lead generation to actually closing sales for you**, AI can guide each step, ensuring your efforts are directed where they have the greatest impact. Incorporate AI to refine your sales funnel, creating a seamless flow that moves prospects effortlessly towards a purchase (or more).

Sales Forecasting for Strategic Planning and Growth

Finally, AI's predictive power extends into the realm of sales forecasting, offering a prophetic glance into future trends and outcomes.

This foresight allows for prudent planning and strategic adjustments, mirroring Joseph's preparation for the years of famine in Egypt. Utilize AI for accurate sales forecasting, providing your business with the invaluable gift of foresight, **enabling you to plan strategically for both abundance and scarcity**.

The bottom line for entrepreneurs and small business owners is clear: to sit at the table of global commerce and feast to one's full satisfaction, **one must be willing to adopt the tools and strategies that level the playing field**. Thus, integrate AI with intent and watch your business flourish.

Through perseverance and faith in the tools provided, small businesses can now embark on a journey that once seemed insurmountable. Adopt AI and harness its capabilities, allowing your business to soar like an eagle amongst the clouds of competition, unhindered by the turbulence below. Take heart, for just as David triumphed over Goliath with a single stone, your small business, armed with the slingshot of AI, can overcome the giants of industry. Be bold, be steadfast, and let AI be your stone in the battle for global competitive clout.

10

THE AI FRONTIER: SEIZING THE OPPORTUNITY NOW

In the amber-glow of a late afternoon, amidst shelves lined with worn business biographies and buzzing gizmos of uncertain purpose, Michael strode through his eclectic storefront. The tiny bells above the door clattered, announcing another potential patron, but he remained engrossed in thought. He pondered scriptures read at dawn, clinging to the promise of mustard-seed faith moving mountains as he faced his own entrepreneurial Mount Everest.

His hands, creased with work and worry, fingered the frayed edge of a faded invoice—symbol of dreams yet poised for flight. "Is it faith or folly," mused Michael, "that bids me summon invisible helpers from this trite circuitry?" He toyed with the idea of AI, as a craftsman might a new chisel, both enticed by its newness and daunted by its strangeness. Somewhere between the clack of keyboards and the hum of servers, he imagined his business, reinvigorated.

A customer, searching for guidance on a perplexing device, ripped Michael from his daydreaming. He assisted with the ease of a man who had spent more time amid cables and codes than with his family, dispensing advice softened by parables of sowing and reaping. "Service birthed this venture," he reflected, "but can intelligence, artificial yet profound, be the manna to sustain it?"

Michael closed the shop as sunset painted the sky in hues of revelation. In the quiet of his own counsel, he resolved to embrace this partner in progress. It was a leap, akin to David before Goliath, wielding not a sling but the slingshot of modernity. The retreat to his study was a pilgrimage, there to prod the keyboard and awaken the dormant algorithms with newfound intent.

The night encroached as he perused the tools - AI for forecasting, prompts for content creation, chatbots for customer service, and algorithms to optimize his supply chain. He was crafting not just a business plan, but an ark to navigate the deluge of change. Each click was a covenant, a melding of his sweat-stained expertise with the cold precision of data.

But how, amidst the rush of transformation and the din of paradigms collapsing, does one keep the essence of one's trade alive? Is it possible that amidst the wires and waves of this digital dawn, the spirit of enterprise and innovation can be not only preserved, but enlivened?

The Time for Action is Now

The entrepreneurial spirit has always thrived on innovation and adaptation. Now, as we stand at the precipice of a new era, **Artificial Intelligence** (AI) stands as the beacon of this transformative age for small businesses. **The moment to harness AI is not in some undefined future; it is right at our fingertips,** ready to be woven into the very fabric of small business growth.

This digital awakening calls for entrepreneurs to be like the wise builders from Matthew 7:24-25, who built their houses on rock instead of sinking sand. With a foundation in AI, businesses can withstand the storms of a competitive market. Luxuries of yesterday have become today's necessities, and AI is no exception. No longer just the domain of corporate giants, AI beckons small businesses with promises of enhanced efficiency, personalized customer experiences, and insights that were once beyond reach.

Embrace AI, Embrace Growth

It is a call to action, resonating deeply within the industrious hearts of entrepreneurs: **embrace AI and unlock the full potential of your business.** Just as David faced Goliath, not with overwhelming force but with a precise and agile approach, small businesses have the advantage of agility, able to integrate AI effectively and swiftly to outmaneuver larger competitors. It is time to elevate your vision and translate your business objectives into a language that AI can act upon, turning myriad data points into a clear strategy for growth.

Envision AI as a diligent collaborator in your business operations, one that does not tire or falter. It can undertake the labor of many — be it in customer service with chatbots, in sales with predictive analytics, or in management with workflow automation. These are the tangible tools AI offers entrepreneurs, equipping them to work smarter and unveiling fresh opportunities. As Proverbs imparts wisdom on seeking counsel, integrating AI can be akin to **adding an infinitely knowledgeable advisor to your team**.

Adopting AI, however, requires a thoughtful approach. One must **balance the adoption of technology with a normative ethical compass**, ensuring that AI serves the business without compromising core values. The same discernment that guides fair business practices should inform AI integration, upholding principles of **integrity and respect** for both users, employees and customers. This balance reflects not only sound strategy but also an adherence to the values that often form the bedrock of small enterprises.

As with any significant business decision, the integration of AI demands education and a measure of courage. Just as Moses encouraged Joshua to be 'strong and courageous' when leading the Israelites to the Promised Land, business owners should educate themselves about AI's potential and take bold steps to incorporate it into their processes. This includes **staying abreast of the latest AI developments** and understanding the relevance to one's specific industry. It is in this continuous learning and application that businesses will thrive.

Therefore, let this serve not as a clarion call but as a sober reminder of the inevitability of change and the pressing need to act judiciously yet decidedly. **The landscape of business evolves ceaselessly, and within it, the entrepreneurs who harness AI stand to become the architects of the new era.** As with the parable of the talents, there is a profound responsibility to invest wisely the resources now at your disposal. AI represents such a resource — a tool to be utilized with foresight and sagacity.

Embrace AI as a Partner in Your Entrepreneurial Quest

When stepping into the realm of entrepreneurship, one must seek partners that magnify our strengths and compensate for our weaknesses. It is written, *"Two are better than one, because they have a good return for their labor"* (Ecclesiastes 4:9). This profound wisdom extends beyond human interactions and finds relevance in our symbiotic relationship with technology. Artificial Intelligence (AI) is not just a tool; it's a **fundamental collaborator** in the modern business world. For the growth-focused entrepreneur, now is the time to embrace AI as the partner that works tirelessly to refine, predict, and enhance every facet of your enterprise. The landscape where small businesses thrive has become fertile grounds for AI integration.

The AI Transformation of Small Business

With AI at your side, the transformation of tedious tasks into automated processes becomes a reachable reality. Consider the time saved when AI algorithms handle inventory management or conduct financial analysis. Reallocating those precious hours enables you to focus on **strategic growth** and personal engagement with clients — core areas that feed the soul of your business. In essence, AI gifts you the time to nurture relationships and **craft your business legacy**, aligning with the spiritual principle of diligent stewardship over the tasks entrusted to us.

Building a Future with AI

Embarking on the AI journey requires vision, courage, and a relentless pursuit of excellence—a combination that defines the quintessential entrepreneur. Remember, the most profound transformations often stem from small, deliberate actions. **Start small, think big, and use AI to unfold the vast potential of your enterprise**. As you cultivate this partnership with AI, may you receive divine favor upon your wise embrace of the opportunities AI grants to today's growth-minded entrepreneurs and small business owners.

Educate yourself about the fundamental concepts. I recall the early days of my own business, sifting through countless sources to distill the essence of AI. Learning the basics through free online courses, webinars, or even AI-focused podcasts is vitally important. This knowledge forms the bedrock of your AI strategy, helping you ask the right questions and make informed decisions.

Evaluate your business needs. Take a moment to reflect on the areas of your business that could benefit from AI. Whether it be customer service, sales forecasting, or inventory management, identifying specific areas for AI intervention can prove transformative. A mindful assessment of your business's health is key to understanding where AI can make a positive impact.

Step into the AI Arena

Once the groundwork is laid, it's time to **experiment with AI tools**. There are a plethora of AI tools developed with small businesses in mind. For example, at the time of writing this book, *Chatbots* such as Drift or Intercom can handle customer inquiries, freeing you to focus on other tasks. Tools like *Zendesk* use AI to enhance customer support, while *Crisp* aggregates customer data to offer personalized services.

There is wisdom in the counsel of many. **Seek advice** from peers who have integrated AI into their businesses. Their experiences, whether marked by success or learning curves, will offer invaluable insight. Moreover, collaborating with AI vendors and attending industry events can provide deeper understanding and personalized guidance tailored to your specific business context.

The Financial Investment of AI

Venturing into AI does require an investment of resources, but consider it not as an expense but as planting seeds for a bountiful harvest. The scripture shares in Ecclesiastes 11:6, *"In the morning sow your seed, and in the evening do not withhold your hand."* With AI, you are sowing into the operational efficiency and future profitability of your business. Start small, using cost-effective AI tools, and as you scale, so too can your investment in more advanced AI solutions.

Streamline with Automation

Automating routine tasks is a stepping stone in AI adoption. Apps like *Clockwise* can manage scheduling, while *QuickBooks* employs AI for financial transactions and record-keeping. As you grow familiar with these applications, your confidence in AI's benefits will solidify, leading to further exploration of AI's capabilities.

Data-Driven Decisions with AI

Harness **data analytics** tools to extract powerful business intelligence. Platforms such as *Tableau* and *Microsoft Power BI* analyze vast amounts of data, delivering insights that can drive your business strategy. Remember the parable of the talents; it teaches us to be good stewards of what we've been given. In the context of AI, this translates to prudently using the data at your disposal for the growth and betterment of your enterprise.

Dive deep into **customer experience enhancement**. Personalized marketing campaigns and improved customer service are within reach through AI tools like *Klaviyo, HighLevel* and *HubSpot*. Their AI-driven insights can lead to increased customer satisfaction and loyalty, which in turn can result in a thriving business.

Safeguarding Your Ethical Compass

In this quest to integrate AI into your business, **maintain an ethical compass**. Ensure the AI tools you choose to abide by privacy standards and ethical guidelines. The moral imperative is clear, and as stewards of new technology, we too must uphold integrity, protecting the trust customers place in our businesses.

In conclusion, as entrepreneurs, we have been granted the serendipitous opportunity to steer the ship of our business into thriving waters with the help of AI. These practical steps, along with a guided moral compass, lay the foundation for a future where our dreams of growth and innovation are realized, honoring both our business ambitions and our spiritual values.

Seizing the AI Opportunity

AI is not the future; it is the present. It is the Daniel in the den of competitors, equipped with the foresight to navigate challenges and emerge unscathed. Your business is the loaves and fishes that, with the right multiplication - this time through AI - can feed the multitudes. It is through this prism we must view our ventures, recognizing in AI an abundant resource entrusted to our care to foster growth and prosperity.

Practical Wisdom and Spiritual Wealth

It has been wisely said that without vision, the people perish. Within the sphere of business, this vision translates into the practical steps and strategies this book has outlined—your map to a promised land that flows with the milk and honey of efficiency, sales, and customer satisfaction. Harnessing AI represents a modern-day stone for David's sling, equipping small business owners with the means to topple Goliath-sized challenges.

The Virtue of Perseverance in Entrepreneurship

The entrepreneurial journey is akin to the construction of Solomon's temple, a labor of devotion where every stone is laid with purpose. Just as the temple could not be built in a day, neither is the integration of AI an instant gratification. Ensure that the foundation is strong, the architecture sound, and the adornments reflect the uniqueness of your mission.

Your call to action is clear: **Rise and begin building. Take this wealth of knowledge and convert it into action**. Empower your enterprise with AI tools that will fortify your operations, just as wisdom fortifies the soul, and watch as growth and prosperity follow suit.

Remember, wisdom is vindicated by her deeds. The integration of AI is your prudent act. It is your Esther moment, for such a time as this, to **make decisions that will echo through the financial statements and customer testimonials of your future**. Remember that the race is not to the swift, nor the battle to the strong, but to those who persevere. In the pursuit of AI, let perseverance be your ally, innovation your chariot, and faith your compass.

The Future Written Today

Today's decisions are tomorrow's legacy. Let the rich tapestry of your business's story be woven with threads of innovation, marked by the integration of AI. With every word you write in this chapter of your journey, you shape the narrative of success that others will read and learn from in times to come. By bridging the tangible with the spiritual, this AI endeavor ascends from mere strategy to a calling – yours to heed, yours to fulfill.

Your mission, should you choose to accept it, begins now. Embark on your AI journey with the full armor of wisdom, the staff of perseverance, and the crown of success surely within your reach.

Embarking on a Journey Beyond the Horizon

As we reach the closing pages of our exploration into the transformative power of Artificial Intelligence (AI) for small business growth and ingenuity, let us pause and reflect upon the multitude of ways this journey has equipped us to navigate the exhilarating landscape that lies ahead. In a world where the tides of technology ebb and flow with relentless vigor, we are called to be the steadfast navigators of these waters, charting a course toward growth and prosperity for our enterprises within the bountiful seas of possibility.

Throughout these chapters, we have uncovered the **potent potential of AI** to revolutionize the small business realm. Together, we ventured through the intricacies of AI integration, unlocking the secrets to **amplifying sales**, **outflanking competitors**, and nurturing your business into the flourishing dream enterprise you envision — all without the apprehension of prohibitive costs.

In your own business ventures, may the wisdom from this trove of knowledge serve as a beacon. **Employ AI to refine customer experiences**, **optimize operational efficiency**, and **harness data-driven decision-making**. Embrace this digital renaissance, and empower your business to scale new heights in the marketplace with precision and grace.

Harnessing AI's Power in the Entrepreneurial Spirit

As we recapitulate the essence of our discussion, remember that AI, in its myriad forms, is a tool that can harmonize seamlessly with the human touch. **AI augments our capabilities, making us better marketers, more insightful salespeople, and more strategic business developers**. AI can transform customer interactions into personalized experiences, elevate product recommendations to an art form, and turn vast oceans of data into rivers of actionable wisdom.

As a proponent of integrating faith and work, use the principles of stewardship as your guide. Like the faithful servant in the Parable of the Talents, invest in the talents of AI technology you've been given to yield an abundant harvest for your business.

Knowing Your Path and Walking It Boldly

I encourage you to step forward with fortitude, armed with the knowledge we've shared. Craft a strategic AI implementation plan, one that aligns with your unique business goals and values. Small changes can yield remarkable results. **Begin with one AI-enabled service or tool; learn its nuances, measure its impact, and then build from there.** This incremental approach not only mitigates risk but also allows for a deeper understanding and appreciation of the technology.

Be mindful of the limitations we have acknowledged—the nascent nature of AI means that the landscape is ever-changing and requires continuous learning and adaptation. As entrepreneurs, we must remain attentive to emerging trends and **maintain a posture of perpetual studenthood** in this field.

Answer the Call to Action

Let this book be your clarion call — to dare the unknown, to innovate boldly, and to lead your enterprise into the dawn of the AI revolution with the assurance of success. For the spirit that beckons to us is one of relentless ambition and boundless optimism, every challenge faced is an opportunity for growth and a testament to the grace and greatness of our purpose on this earth.

The AI revolution in small business is not just an abstract concept; it is **a tangible transformation waiting to be realized by those with the faith to believe and the courage to act**. Invoke the wisdom of the past and the wonders of the present to sculpt a future that beckons with opportunity and resonates with your ultimate mission to grow your people, your business, and His Kingdom.

In the words of a man who understood the profound impact of technology on society, let us take inspiration as we venture forth:

"We're still in the first minutes of the first day of the Internet revolution." – Scott Cook

May your entrepreneurial journey through the ever-evolving world of AI be as inspiring and revolutionary as the internet has been for humanity. Go, grow, and let the AI revolution uplift your aspirations to heavenly heights to makeover your small business into your dream business!

ABOUT THE AUTHOR

What is most important to Philip Blackett and what truly forms his identity is his relationship with his Lord and Savior Jesus Christ. Philip's mission for the rest of his life is to Grow God's People, Grow God's Businesses, and Grow God's Kingdom as a good and faithful steward of all God has entrusted him, while having a positive influence on all who he encounters each day as a Kingdom Man.

Professionally speaking, Philip is passionate about helping entrepreneurs and small business owners grow their dream businesses, while utilizing his skillset in sales, marketing and business development. Previously, Philip served as President of Cemetery Services, Inc., a seven-figure business he bought based in the Greater Boston area. It was "his pleasure" to also serve as a Manager for a Chick-Fil-A restaurant.

At FedEx, Philip previously provided support to several senior Marketing executives (including the current CEO) as a Senior Communications Specialist after working on its Corporate Social Responsibility team. Before FedEx, Philip advised investors on Wall Street in New York City as an Equity Research Analyst for Goldman Sachs, where he helped recommend investments in over 100 publicly traded companies across ten industries.

Regarding his education, Philip graduated from the Southern Baptist Theological Seminary with his Masters of Divinity (M.Div) degree with a concentration in Great Commission Studies. He also earned his MBA from Harvard Business School. In college, Philip graduated from the University of North Carolina at Chapel Hill as a Morehead-Cain Scholar, majoring in Political Science and Economics.

Philip is a Life Member of Alpha Phi Alpha Fraternity, Inc. When he is not actively fulfilling his mission, Philip enjoys reading, watching sports, and raising his twin daughters, Sofia and Elizabeth, with his wife Mayra.

— · —

BOOKS BY PHILIP

Disagree without Disrespect: How to Respectfully Debate with Those who Think, Believe and Vote Differently than You

Future-Proof: How to Adopt and Master Artificial Intelligence (A.I.) to Secure Your Job and Career

The Unfair Advantage: How Small Business Owners can Use Artificial Intelligence (A.I.) to Boost Sales, Outsmart the Competition and Grow their Dream Businesses without Breaking the Bank

Jesus over Black: How My Faith Transformed Me into a Conservative within the Black Community

Maverick Lineage: What I Learned about Black Conservatism in America

Bridging the GOP Gap: How the Republican Party can Win Over African American Voters with Inclusivity and Trust without Compromising Values

CONNECT WITH PHILIP

f

facebook.com/PhilipBlackettFB

🐦

twitter.com/PhilipBlackett

in

linkedin.com/in/philipblackett

📷

instagram.com/philipblackett

▶

youtube.com/@PhilipBlackett

♪

tiktok.com/@pblackett

Facebook:

https://www.facebook.com/PhilipBlackettFB

X (Twitter):

https://twitter.com/PhilipBlackett

LinkedIn:

https://www.linkedin.com/in/philipblackett

Instagram:

https://instagram.com/philipblackett

YouTube:

https://www.youtube.com/@PhilipBlackett

TikTok:

https://www.tiktok.com/@pblackett

Blog:

https://www.PhilipBlackett.com